S. Kusuoka, T. Maruyama (Eds.)

Advances in Mathematical Economics

Volume 4

Springer

Shigeo Kusuoka
Professor
Graduate School of Mathematical Sciences
University of Tokyo
3-8-1 Komaba, Meguro-ku
Tokyo, 153-0041 Japan

Toru Maruyama
Professor
Department of Economics
Keio University
2-15-45 Mita, Minato-ku
Tokyo, 108-8345 Japan

ISBN 978-4-431-99821-1

Printed on acid-free paper
Springer-Verlag is a company in the BertelsmannSpringer publishing group.

Printed in Japan

Advances in MATHEMATICAL ECONOMICS

Aims and Scope. The project is to publish *Advances in Mathematical Economics* once a year under the auspices of the Research Center of Mathematical Economics. It is designed to bring together those mathematicians who are seriously interested in obtaining new challenging stimuli from economic theories and those economists who are seeking effective mathematical tools for their research.

The scope of *Advances in Mathematical Economics* includes, but is not limited to, the following fields:

- Economic theories in various fields based on rigorous mathematical reasoning.
- Mathematical methods (e.g., analysis, algebra, geometry, probability) motivated by economic theories.
- Mathematical results of potential relevance to economic theory.
- Historical study of mathematical economics.

Authors are asked to develop their original results as fully as possible and also to give a clear-cut expository overview of the problem under discussion. Consequently, we will also invite articles which might be considered too long for publication in journals.

Springer
Tokyo
Berlin
Heidelberg
New York
Barcelona
Hong Kong
London
Milan
Paris
Singapore

Table of Contents

Adv. Math. Econ. 4, 1–23 (2002)

Advances in
MATHEMATICAL
ECONOMICS

Abstract convexity and nonsmooth analysis: Global aspects*

A.D. Ioffe[1] **and A.M. Rubinov**[2]

[1] Department of Mathematics, Technion, Haifa 32000, Israel
[2] School of Information Technology and Mathematical Sciences, University of Ballarat, Victoria 3353, Australia

Received: June 7, 2001
Revised: September 13, 2001

JEL classification: C61

Mathematics Subject Classification (2000): 46 E 30, 28 A 20, 60 B 12

Abstract. We study some types of convergence of abstract convex functions. We show that under certain conditions, convergence of support sets of a sequence of abstract convex functions implies graph convergence of the support sets at a point and the global subdifferential. Some examples and counterexamples are given

Key words: Abstract convexity, support sets, subdifferential, convergence

1. Introduction

Models of abstract convexity naturally arise in certain applications, in particular, connected with economics (theory of production functions and utility functions). They are also very helpful for theoretical study of some problems of global optimization (see, for example, [12], Chapter 8) and for numerical methods of global optimization ([12], Chapter 9). The main technical complications in studying models of abstract convexity are caused by the fact that they often lack two basic technical advantages of the classical convexity: the equivalence of the local and global characterizations of subdifferentials and the equivalence of the Fenchel and Minkowski types of duality.

Let us be a bit more specific, first about the latter. Let f be a convex function defined on a Banach space X and let X^* be the conjugate

* This research has been supported by Australian Research Council.

space to X. Let $x \in X$. According to the Fenchel duality scheme, we associate with f its *conjugate* $f^*(x^*) = \sup_x(\langle x^*, x\rangle - f(x))$. In the Minkowski duality (usually applied to sublinear functions) we associate with f its *support set* which is the collection of affine functions $h(x)$ satisfying $f(x) \geq h(x)$ for all x. The differential properties of f at a given x in the first scheme are described by means of the *subdifferential* $\partial f(x)$of f at x which is defined as the collection of x^* satisfying $f(y) - f(x) \geq \langle x^*, y - x\rangle$ for all x. In the second model a similar role is played by the *support* $\partial^* f(x)$ of f at x which is the collection of all affine functions $h(x) = \langle x^*, x\rangle - c$ ($c \in \mathbb{R}$ – we write for simplicity $h = (x^*, c)$ in what follows and identify the set of affine functions with the product $X^* \times \mathbb{R}$) belonging to the support set of f and satisfying $f(x) = h(x)$ for the given x. Clearly $h = (x^*, c)$ belongs to the support of f if and only if it belongs to the epigraph of f^*, and $x^* \in \partial f(x)$ if and only if $(x^*, c) \in \partial^* f(x)$ where $c = \langle x^*, x\rangle - f(x)$. In other words, there is a simple one-to-one correspondence between the models.

Abstract convexity ([11, 12, 13] forms a convenient framework for generalization of the concept of a support set to broad classes of functions and, through it, to a generalization of the global concept of a subdifferential in those models of abstract convexity in which we may have an analogue of "linear" functions of which "affine functions" can be obtained by adding arbitrary constants. However, when the original set of functions is of a neither type, no suitable Fenchel type duality can be produced and only the Minkowski duality scheme associated with the concept of so called $H-$ *convexity* is available.

Questions about convergence of differential characteristics under various types of convergences of functions attracted considerable attention in convex and local non-smooth analysis since such characteristics play the fundamental role in characterizations and practical search of solutions of optimization problems involving the converging and the limiting functions. In this paper we study convergence of support sets and globally defined subdifferentials in models of abstract convexity. The paper is closely connected with a recent paper by the first author [6] mainly devoted to local aspects of the problem. Here we concentrate on global aspects of abstract convexity.

The main result of the paper proved in Section 3 shows, roughly speaking, that, under certain conditions, convergence of support sets of a sequence of H-convex functions implies graph convergence of the set-valued mapping ∂^*. This theorem contains as a special case the corresponding theorem for global subdifferentials of [6] (and actually leads to a strengthening of the latter) which in turn is an extension of the famous Attouch-Beer theorem [1] connecting slice convergence of convex functions with graph convergence of their subdifferentials. We then consider several applications of the theorem, in particular, in more details, to those models in which global subdifferentials make sense and study convergence of subdifferentials of so-called ICAR functions on the positive orthant which often appear in models of economic growth. In the concluding section we describe some links between certain important concepts of the economic theory (production, utility, cost functions) and the mathematical objects studied in the paper. We hope to present elsewhere a more detailed study of implications of the here obtained results to problems of approximation and stability in certain models of mathematical economics.

2. Main Definitions

Let H be a set of functions defined on a set X. Denote by F_X the union of the set of all functions $f : X \to \bar{\mathbb{R}}_+ := \mathbb{R} \cup \{+\infty\}$ and the function $-\infty$. Let $f \in F_X$. The set $\operatorname{supp}(f, H) = \{h \in H : h \le f\}$ is called the support set of the function f with respect to H. ($h \le f$ means $h(x) \le f(x)$ for all $x \in X$.) A function $f \in F_X$ is called abstract convex with respect to H (or H-convex) if $f(x) = \sup\{h(x) : h \in \operatorname{supp}(f, H)\}$. (As usual, the supremum over the empty set is equal to $-\infty$.) We denote the set of all H-convex functions by $P(H, X)$. Let $f \in P(H, X)$. A function $h \in \operatorname{supp}(f, H)$ is called a *support* to f at a point x if $h(x) = f(x)$. Let $\varepsilon > 0$. A function $h \in \operatorname{supp}(f, H)$ is called ε-support at x if $f(x) \le h(x) + \varepsilon$. We denote the set of all supports (ε-supports) to f at x by $\partial^*_H f(x)$ ($\partial^*_{H,\varepsilon} f(x)$, respectively). For an $f \in P(H, X)$ the set $\partial^*_{H,\varepsilon} f(x)$ is always nonempty, however $\partial^*_H f(x)$ can be empty for all x which are not global minimizers. (If H contains constants and $\bar{x}$ is a global minimizer of f then $f(\bar{x})\mathbf{1} \in \partial^*_H f(\bar{x})$. Here $\mathbf{1}(x) \equiv 1$.) We now give a simple example of a situation, where $\partial^*_H f(x)$ is empty.

Example 2.1. Let $X = \mathbb{R}_+ := \{x \in \mathbb{R} : x \ge 0\}$ and H be the set of all two-step functions $h_{a,b}$ with $a, b \in \mathbb{R}_+$, where

$$h_{a,b}(x) = \begin{cases} 0 & x \le a \\ b & x > a. \end{cases}$$

Assume that H is equipped with the uniform metric: $d(h, h') = \max_x |h(x) - h'(x)|$. It is easy to check that a function $f : \mathbb{R}_+ \to \bar{\mathbb{R}}_+ := \mathbb{R} \cup \{+\infty\}$ is H-convex if and only if $f(0) = 0$, f is increasing ($x \ge y \Longrightarrow f(x) \ge f(y)$) and lower semicontinuous. Let f be a strictly increasing ($x > y \Longrightarrow f(x) > f(y)$) continuous function with $f(0) = 0$. Then the set $\partial^*_H f(x)$ is empty for all $x > 0$.

Consider a metric space H of functions defined on a metric space X. For $f \in P(H, X)$ consider sets:

$$\tilde{\partial}^*_H f = \{(x, f(x), h) \in X \times \mathbb{R} \times H : h \in \partial^*_H f(x)\} \tag{2.1}$$

$$\tilde{\partial}^*_{H,\varepsilon} f = \{(x, f(x), h) \in X \times \mathbb{R} \times H : h \in \partial^*_{H,\varepsilon} f(x)\} \tag{2.2}$$

which will be called the *support jet* (resp. *support ε-jet*) of f. In the sequel an important role will be played by restriction of sets (2.1) and (2.2) to balls. Let $\delta > 0$ and $(x, f(x), h) \in \tilde{\partial}^*_H f$. Set

$$\tilde{\partial}^*_H f(\delta; x, h)$$
$$= \{(x', f(x'), h') \in \tilde{\partial}^*_H f : d(x, x') \le \delta,\ d(h, h') \le \delta,\ |f(x') - f(x)| \le \delta\}$$

and

$$\tilde{\partial}^*_{H,\varepsilon} f(\delta; x, h)$$
$$= \{(x', f(x'), h') \in \tilde{\partial}^*_{H,\varepsilon} f : d(x, x') \le \delta,\ d(h, h') \le \delta,\ |f(x') - f(x)| \le \delta\}. \tag{2.3}$$

If $\varepsilon_1 > \varepsilon_2 > 0$ then

$$\tilde{\partial}^*_{H,\varepsilon_1} f(\delta; x, h) \supset \tilde{\partial}^*_{H,\varepsilon_2} f(\delta; x, h) \supset \tilde{\partial}^*_H f(\delta; x, h). \tag{2.4}$$

We also have

$$\bigcap_{\varepsilon>0} \tilde{\partial}^*_{H,\varepsilon} f(\delta; x, h) = \tilde{\partial}^*_H f(\delta; x, h) \tag{2.5}$$

Assume that the space $X \times \mathbb{R} \times H$ equipped with the metric $d((x', c', h'), (x, c, h)) = d(x', x) + |c' - c| + d(h', h)$ and the totality of all bounded nonempty subsets of this space is equipped with the Hausdorff metric. Conditions, which insure that

$$\tilde{\partial}^*_{H,\varepsilon} f(\delta; x, h) \to \tilde{\partial}^*_H f(\delta; x, h) \ \text{ as } \varepsilon \to +0. \tag{2.6}$$

is of certain interest. Due to (2.4)), we conclude that (2.6) is equivalent to the following: for each $\eta > 0$ an $\varepsilon > 0$ can be found which has the following property: for every $(x_*, f(x_*), h_*) \in \tilde{\partial}^*_{H,\varepsilon} f(\delta; x, h)$ there exists $(\hat{x}, f(\hat{x}), \hat{h}) \in \tilde{\partial}^*_H f(\delta; x, h)$ such that $d(x_*, \hat{x}) \le \eta$, $|f(x_*) - f(\hat{x})| \le \eta$ and $d(h_*, \hat{h}) \le \eta$.

If X, H are locally compact spaces and f is a continuous function, then sets $\tilde{\partial}^*_{H,\varepsilon} f(\delta; x, h)$ are compact. Applying (2.4) and (2.5) we conclude that (2.6) holds. However convergence (2.6) often holds without local compactness. At the same time Example 2.1 demonstrates that in general local compactness can not be omitted.

Let $h \in \partial^*_H f(x)$ and $\delta > 0$. Consider the function

$$r_{f,\delta;x,h}(\varepsilon) = d(\tilde{\partial}^*_{H,\varepsilon} f(\delta; x, h), \tilde{\partial}^*_H f(\delta; x, h)) \tag{2.7}$$

defined on $\mathbb{R}_+$. Clearly (2.6) is equivalent to $r_{f,\delta,x,h}(\varepsilon) \to 0$ as $\varepsilon \to +0$.

Recall the definition of Γ-convergence (see e.g. [4]). Let X be a metric space. A sequence of functions $f_n : X \to [-\infty, +\infty]$ is said to Γ-converge to a function f if

$$\liminf f_n(x_n) \geq f(x) \text{ for each } x \in X \text{ and } x_n \to x \tag{2.8}$$

and

$$\text{for each } x \in X$$
$$\text{there exists a sequence } x_n \to x \text{ such that } f_n(x_n) \to f(x). \tag{2.9}$$

We now introduce a special convergence for H convex functions. Let H be a metric space of functions defined on a metric space X.

Definition 2.1. We say that a sequence of H-convex functions (f_n) H-converges to an H-convex function f (briefly $f_n \xrightarrow{H} f$) if (2.9) holds and

$$\text{supp}\,(f, H) \subset \text{Li supp}\,(f_n, H), \tag{2.10}$$

Here *Li* stands for the *lower* limit: for each $h \in \text{supp}\,(f, H)$ there exists a sequence $h_n \in \text{supp}\,(f_n, H)$ such that $h_n \to h$.

Let us give a simple example of H-convergence. Let $f_n, f \in P(H, X)$ $f_n \geq f$ and $f_n(x) \to f(x)$ for all $x \in X$. Then $f_n \xrightarrow{H} f$. Assume that $h \in H \Longrightarrow h - c\mathbf{1} \in H$ for all $c \in \mathbb{R}_+$ and assume that $f_n(x) \to f(x)$ for all $x \in X$ and there is a sequence $\varepsilon_n \to 0$ such that $f_n \geq f - \varepsilon_n \mathbf{1}$ for all n. Then $f_n \xrightarrow{H} f$.

We now present an example which demonstrates that in general an H-convergent sequence can have more than one limit.

Example 2.2. Let $X = \{1, 1/2, \dots, 1/n, \dots; 0\}$ and let H be the set of all functions defined on X. Assume that H equipped with the metric d:

$$d(h, h') = |h(0) - h'(0)| + \sum_{n=1}^{+\infty} 2^{-n}(h(1/n) - h'(1/n)).$$

Let

$$f_n(x) = \begin{cases} 0 & x = 1/n \\ 1 & x \in X,\ x \neq 1/n. \end{cases}$$

Let, further

$$f(x) = \begin{cases} 0 & x = 0 \\ 1 & x \in X,\ x \neq 0. \end{cases}$$

Clearly for each $x \in X$ there exists a sequence $x_n \to x$ such that $x_n \to x$ and $f(x_n) \to f(x)$. Let $h \in \operatorname{supp}(f, H)$, that is, $h(0) \leq 0$, $h(x) \leq 1$ for $x \neq 0$. Let

$$h_n(x) = \begin{cases} h(0) & x = 0 \\ \min(h(x), f_n(x)) & x \neq 0. \end{cases}$$

Then $h_n \in \operatorname{supp}(f, H)$ and $h_n \to h$, so $f_n \xrightarrow{H} f$. Consider now the function $g = \mathbf{1}$. It is easy to check that $f_n \xrightarrow{H} g$ as well.

To avoid such unpleasant situations, we need to impose some restrictions on the possible choice of functions included into H.

Definition 2.2. We say that H is a match for X or just that X and H match if $h_n \to h$ and $x_n \to x$ implies $h_n(x_n) \to h(x)$.

Proposition 2.1. *If X and H are metric spaces and H is a match for X, then H-convergence implies Γ-convergence.*

Proof. Let $f_n \xrightarrow{H} f$. We need only to check that $\liminf_{n \to +\infty} f_n(x_n) \geq f(x)$ for arbitrary $x \in X$ and any sequence $x_n \to x$. Let ε be an arbitrary positive number and $h \in \partial^*_{H,\varepsilon} f(x)$. It follows from (2.10) that there exists a sequence $h_n \to h$ such that $h_n \in \operatorname{supp}(f_n, H)$. Since $h_n(x_n) \leq f_n(x_n)$, and X and H match, we have $h_n(x_n) \to h(x) \geq f(x) - \varepsilon$, so $\liminf_n f_n(x_n) \geq f(x) - \varepsilon$. □

Corollary 2.1. *If X and H match, then each H-convergent sequence has an unique limit.*

3. Main Theorem

Despite Proposition 2.1, the approximation provided by H-converging sequences is rather rough. Much more interesting would be convergence of local supports. There is no hope for a pointwise convergence of supports at every point under reasonable conditions. This does not happen even in the classical situation of convex functions (in the usual sense). The relation $\tilde{\partial}^*_H f \subset \operatorname{Li} \tilde{\partial}^*_H f_n$ could be a reasonable compromise. In order to guarantee this type of convergence we need to strengthen the definition of H-convergence.

Definition 3.1. *We say that f_n strongly H-converges to f ($f_n \stackrel{sH}{\to} f$) if $f_n \stackrel{H}{\to} f$ and for each $(x,h) \in \tilde{\partial}^*_H f$ there exists $\delta > 0$ such that*

$$r_{f_n,\delta;x,h}(\varepsilon) \to 0 \text{ as } n \to +\infty \text{ and } \varepsilon \to +0. \tag{3.1}$$

where $r_{f,\delta;x,h}(\varepsilon)$ is defined by (2.7).

Theorem 3.1. *Let metric spaces X and H match. Let a sequence of H-convex functions f_n strongly H-converges to an H-convex function f. Then*

$$\tilde{\partial}^*_H f \subset \operatorname{Li} \tilde{\partial}^*_H f_n. \tag{3.2}$$

Proof. Let $h \in \partial^*_H f(x)$. By Proposition 2.1 there exists a sequence $x_n \in X$ such that $x_n \to x$ and $f_n(x_n) \to f(x)$. Applying (2.10) and the inclusion $h \in \operatorname{supp}(f,H)$, we conclude that there exists a sequence $h_n \in \operatorname{supp}(f_n,H)$ such that $h_n \to h$. We have

$$h_n(x_n) \le f_n(x_n), \quad h_n(x_n) \to h(x) = f(x), \quad f_n(x_n) \to f(x).$$

Let $\varepsilon_n = f_n(x_n) - h_n(x_n) \ge 0$. Then $\varepsilon_n \to +0$. We also have $h_n \in \partial^*_{H,\varepsilon_n} f_n(x_n)$, that is $(x_n, f_n(x_n), h_n) \in \tilde{\partial}^*_{H,\varepsilon_n} f_n$. Let $\delta > 0$ be a number such that (3.1) holds at the point (x,h). Then $(x_n, f_n(x_n), h_n) \in \tilde{\partial}_{H,\varepsilon_n}(f_n,\delta;x,h)$ for sufficiently large numbers n. Hence there exists $(\hat{x}_n, f(\hat{x}_n), \hat{h}_n) \in \tilde{\partial}^*_H(f_n,\delta;x,h))$ such that

$$d(x_n,\hat{x}_n) \le r_{f_n,\delta;x,h}(\varepsilon_n), \quad |f_n(x_n) - f_n(\hat{x}_n)| \le r_{f_n,\delta;x,h}(\varepsilon_n)$$
$$d(h_n,\hat{h}_n) \le r_{f_n,\delta;x,h}(\varepsilon_n).$$

Since $x_n \to x$, $h_n \to h$ and $f_n(x_n) \to f(x)$ it follows from (3.1) that $\hat{x}_n \to x$, $\hat{h}_n \to h$ and $f(\hat{x}_n) \to f(x)$. □

The following example presents a simple situation in which the theorem works. Later in Section 6 we shall show that the H-convergence alone does not imply the convergence of support jets.

Example 3.1. Let X be a metric space, and let H consist of all functions of the form $h(x) = \beta - \alpha d(x,y)$ with $\alpha \in \mathbb{R}_+$, $\beta \in \mathbb{R}$, $y \in X$. Then f is H-convex if and only if it is lower semicontinuous, bounded below on bounded sets and satisfies

$$\liminf_{d(x,y)\to\infty} \frac{f(x)}{d(x,y)} > -\infty, \ \forall y,$$

in case when the diameter of X is infinite. (A set is bounded if its diameter is finite. Observe also that the lim inf does not actually depend on y.)

Suppose now that we have a sequence (f_n) of H-convex functions converging to a certain function f in the following sense: (2.9) is satisfied (e.g. f_n pointwise converge to f) and

$$\liminf_{n\to\infty} \inf_x \frac{f_n(x) - f(x)}{d(x,y)+1} = 0. \tag{3.3}$$

The latter condition means that f_n cannot decrease at infinity much faster than f. Observe again that the validity of (3.3) does not depend on the choice of y. Moreover, if it holds then

$$\liminf_{n\to\infty} \inf_x \frac{f_n(x) - f(x)}{d(x,z)+\varepsilon} = 0.$$

for any $z \in X$ and any $\varepsilon > 0$. Indeed, (3.3) is equivalent to

$$\limsup_{n\to\infty} \sup_x \frac{(f_n(x) - f(x))^-}{d(x,y)+1} = 0$$

(where $\alpha^- = \max\{-\alpha, 0\}$). Therefore

$$\begin{aligned} 0 &\le \limsup_{n\to\infty} \sup_x \frac{(f_n(x) - f(x))^-}{d(x,z)+\varepsilon} \\ &= \limsup_{n\to\infty} \sup_x \frac{(f_n(x) - f(x))^-}{d(x,y)+1} \cdot \frac{d(x,y)+1}{d(x,z)+\varepsilon} \\ &\le \frac{d(x,y)+1}{\varepsilon} \limsup_{n\to\infty} \sup_x \frac{(f_n(x) - f(x))^-}{d(x,y)+1} = 0. \end{aligned}$$

It is an easy matter to see that (f_n) H-converges to f. Indeed, assume that $f(x) \ge \beta - \alpha d(x,y)$ for all x. Take an $\varepsilon > 0$ and choose $\bar{n}$ so large that $f_n(y) \ge f(y) - \varepsilon$ and $f_n(x) - f(x) \ge -\varepsilon(d(x,y)+1)$ for all x if $n \ge \bar{n}$. Then $f_n(x) \ge (\beta - 2\varepsilon) - (\alpha + \varepsilon)d(x,y)$ and H-convergence follows.

On the other hand, if we consider H with the natural product metric of $\mathbb{R} \times \mathbb{R} \times X$, then H clearly becomes a match for X. Therefore by the theorem, for any x and any $h \in \partial_H^* f(x)$ there are sequences $(x_n) \to x$ and $(h_n) \to h$ such that $f(x_n) \to f(x)$ and $h_n \in \partial_H^* f_n(x_n)$.

4. H-affine functions and ω-convergence

Let (X, L) be a pair of sets coupled by a function $\omega : L \times X \to \mathbb{R}$. We consider an element $l \in L$ as a function defined on X by $l(x) = \omega(l, x)$ and an element $x \in X$ as a function defined on L by $x(l) = \omega(l, x)$. Let $l \in L$ and $c \in \mathbb{R}$. Denote by (l, c) the function $h = l - c\mathbf{1}$, where $\mathbf{1}(x) \equiv 1$ for all $x \in X$. In other words $(l, c)(x) = \omega(l, x) - c$ $(x \in X)$. We shall call functions (l, c) L-affine functions. Let $f \in F_X$. The function f^ω defined on L by $f^\omega(l) = \sup_{x \in X}(\omega(l, x) - f(x))$ is called ω-conjugate to f. The following result is well-known (see, for example [12]) and easily follows from definitions.

Proposition 4.1. *Let $f \in F_X$. Then* $\operatorname{epi} f^\omega = \operatorname{supp}(f, H)$, *where H is the set of all L-affine functions and* $\operatorname{epi} f^\omega = \{(l, c) : c \geq f^\omega(l)\}$ *is the epigraph of the function f^ω. In particular, f is H-convex if and only if* $f(x) = \sup_l(\omega(l, x) - f^\omega(l))$.

In this section we consider H-convex functions with H being the collection of L-affine functions. It is clear that H is a match for X if and only if the coupling $\omega(l, x)$ is a continuous function. A natural convergence for H-convex functions of this type (extending the slice convergence of standard convex functions) was introduced in [6] under the name of $\omega - convergence$. Recall the definition. Let L be a metric space of functions defined on a metric space X. Assume that L and X match, that is the coupling function $\omega(l, x) := l(x)$ is continuous. Let H be the set of all L-affine functions $h(x) = l(x) - c$. Then H can be naturally interpreted as the product of L and $\mathbb{R}$; accordingly, we consider H with the metric $d((l, c), (l', c')) = d(l, l') + |c - c'|$.

Definition 4.1. ([6]) Let $f_n \in P(H, X)$, $n = 1, 2\ldots$ and $f \in P(H, X)$. Then (see ([6]) $f_n \to_\omega f$ if

$$\forall\ x \in X\ \exists\ x_n \to x : \ \lim f_n(x_n) = f(x); \tag{4.1}$$

and

$$\forall\ l \in L\ \exists\ l_n \to l : \ \lim_n f_n^\omega(l_n) = f^\omega(l). \tag{4.2}$$

Proposition 4.2 ([6]). *(f_n) ω-converges to f if and only if (f_n) Γ-converges to f and (f_n^ω) Γ-converges to f^ω.*

Our immediate goal is to establish relationship between H-convergence and ω-convergence for H-convex functions when H is the collection of L-affine functions. Let $h = (l, c)$, $h_n = (l_n, c_n)$. Then of course $h_n \to h$ is equivalent to $l_n \to l$, $c_n \to c$.

Proposition 4.3. *ω-convergence and H-convergence are equivalent on the class of H-convex functions, provided the coupling function $\omega(l,x)$ is continuous.*

Proof. We only need to verify that (4.2) is the same as (2.10), provided (4.1) holds. Assume that (4.2) also holds for a certain sequence of (l_n), and let $h = (l,c) \in \operatorname{supp} f$. By Proposition 4.1 this means that $c \geq f^{\omega}(l)$. It follows from (4.2) that there are $c_n \geq f_n^{\omega}(l_n)$ converging to c. Setting $h_n = (l_n, c_n)$, we see that $h_n \in \operatorname{epi} f_n^{\omega} = \operatorname{supp} f_n$ and $h_n \to h$. This is true for any $h \in \operatorname{supp} f$, whence (2.10).

Conversely, let (2.10) holds together with (4.1). In is enough to verify that (f_n^{ω}) Γ-converge to f^{ω} from which (4.2) immediately follows. Let $l \in L$, and let $x \in X$ be such that either $l(x)-f(x) > f^{\omega}(h)-\varepsilon$, if $f^{\omega}(l) < \infty$ or $l(x) - f(x) > \varepsilon^{-1}$ otherwise. Let $l_n \to l$, and let $x_n \to x$ be such that $f_n(x_n) \to f(x)$. Then $f_n^{\omega}(l_n) \geq l_n(x_n) - f_n(x_n) \to l(x) - f(x)$, so that in either case (whether $f^{\omega}(l)$ is finite or not, $\liminf f_n^{\omega}(l_n) \geq f^{\omega}(l)$.

To prove the "limsup" inequality, take an $\bar{l}$ $\in$dom f^{ω}. Then $h = (l, f^{\omega}(l)) \in \operatorname{epi} f^{\omega} = \operatorname{supp} f$. Take $h_n = (l_n, c_n) \in \operatorname{supp} f_n$ converging to h. Then $c_n \geq f_n^{\omega}(l_n)$ and therefore $\limsup f_n^{\omega}(l_n) \leq f^{\omega}(l)$. □

5. Subdifferentials

Let L be a metric space of functions defined on a metric space X. Let $f \in P(H,X)$, where H is the space of all L-affine functions considered with the natural metric defined in the previous section. A function $l \in L$ is called an L-subgradient of f at a point $x_0 \in X$ if

$$l(x) - l(x_0) \leq f(x) - f(x_0) \quad \text{for all } x \in X.$$

The set $\partial_L f(x_0)$ of all L-subgradients is called the L-subdifferential of f at x_0. Let $\varepsilon \geq 0$. A function $l \in L$ is called an (L,ε)- subgradient of the function f at a point $x_0 \in X$ if

$$l(x) - l(x_0) \leq f(x) - f(x_0) + \varepsilon \quad \text{for all } x \in X.$$

The set $\partial_{L,\varepsilon} f(x_0)$ of all ε-subgradients of f at x_0 is called ε-subdifferential of f at x. Clearly $\partial_{L,0} f(x_0) = \partial_L f(x_0)$.

The following statement is trivial.

Proposition 5.1. *Let $f \in P(H,X)$, $l \in L$, let $c = l(x_0) - f(x_0)$ for a certain $x_0 \in X$, and let $\varepsilon \geq 0$. Then $l \in \partial_{L,\varepsilon} f(x_0)$ if and only if $(l, c+\varepsilon) \in \partial^*_{H,\varepsilon} f(x_0)$.*

Suppose that L and X match, which means that the canonical coupling $\omega(x,l) = l(x)$ is a continuous function. As in the previous section we consider the pair X, H with the coupling function $\bar{\omega}((l,c),x) = (l,x) - c$, and define analogues of sets $\tilde{\partial}^*_H f$ and $\tilde{\partial}^*_{H,e} f$ for L-subdifferentials and (L,ε)-subdifferentials of a function $f \in PH(X)$. Namely, given a $\delta > 0$ and an $l \in \partial_L f(x)$, we set

$$\begin{aligned}\tilde{\partial}_L f(\delta; x, l) = \{&(x', f(x'), l') \in X \times \mathbb{R} \times L : \\ &l' \in \partial_L f(x'),\ d(x,x') \leq \delta,\ d(l,l') \leq \delta,\ |f(x) - f(x')| \leq \delta\}\end{aligned}$$

and

$$\begin{aligned}\tilde{\partial}_{L,\varepsilon} f(\delta; x, l) = \{&(x', f(x'), l') \in X \times \mathbb{R} \times L : \\ &l' \in \partial_{L,\varepsilon} f(x'),\ d(x,x') < \delta,\ d(l,l') < \delta,\ |f(x) - f(x')| \leq \delta\}.\end{aligned}$$

It follows from Proposition 5.1 that

$$(x, f(x), l) \in \tilde{\partial}_L f \Longleftrightarrow (x, f(x), (l, l(x) - f(x)) \in \tilde{\partial}^*_H f; \tag{5.1}$$

$$(x, f(x), l) \in \tilde{\partial}_{L,\varepsilon} f \Longleftrightarrow (x, f(x), (l, l(x) - f(x) + \varepsilon) \in \tilde{\partial}^*_{H,\varepsilon} f. \tag{5.2}$$

For $l \in \partial_L f(x)$ and $\delta > 0$ consider the function

$$s_{f,\delta;x,l}(\varepsilon) = d(\tilde{\partial}_{L,\varepsilon} f(\delta; x, l), \tilde{\partial}_L f(\delta; x, l)).$$

Proposition 5.2. *Let $f \in P(H,X)$, let $l \in \partial_L f(x)$ and $\delta > 0$. Let the function ω be uniformly continuous in the δ-neighborhood of a point (l,x). Consider a function $h \in \partial^*_H f(x)$, where $h = (l,c)$ with $c = l(x) - f(x)$. If $s(\varepsilon) := s_{f,\delta;x,l}(\varepsilon) \to 0$ as $\varepsilon \to +0$ then $r_{f,2\delta;x,h}(\varepsilon) \to 0$ as $\varepsilon \to 0$, where r is a function defined by (2.7).*

Proof. Let $(x_\varepsilon, f(x_\varepsilon), h_\varepsilon) \in \tilde{\partial}^*_{H,\varepsilon} f(\delta, x, h)$. Then $d(x_\varepsilon, x) < \delta$, $d(h_\varepsilon, h) < \delta$ and $h_\varepsilon \in \partial^*_{H,\varepsilon} f(x_\varepsilon)$. We have $h_\varepsilon = (l_\varepsilon, c_\varepsilon + \varepsilon)$, where $l_\varepsilon \in \partial_{L,\varepsilon} f(x_\varepsilon)$ and $c_\varepsilon = l_\varepsilon(x_\varepsilon) - f(x_\varepsilon)$. Since $d(h_\varepsilon, h) = d(l, l_\varepsilon) + |c_\varepsilon + \varepsilon - c| < \delta$ it follows that $d(l, l_\varepsilon) < \delta$. It follows from the definition of $s(\varepsilon) := s_{f,\delta;x,l}(\varepsilon)$ that there exists x'_ε and l'_ε such that $l'_\varepsilon \in \partial_L f(x'_\varepsilon)$ and

$$d(x, x'_\varepsilon) < \delta, \quad d(l, l'_\varepsilon) < \delta; \tag{5.3}$$

$$d(x_\varepsilon, x'_\varepsilon) \leq s(\varepsilon), \quad |f(x_\varepsilon) - f(x'_\varepsilon)| \leq s(\varepsilon), \quad d(l_\varepsilon, l'_\varepsilon) \leq s(\varepsilon). \tag{5.4}$$

Let $h'_\varepsilon = (l'_\varepsilon, c'_\varepsilon)$ where $c'_\varepsilon = l'_\varepsilon(x'_\varepsilon) - f(x'_\varepsilon))$. Then

$$h'_\varepsilon \in \partial^*_H f(x'_\varepsilon) \tag{5.5}$$

and

$$d(h_\varepsilon, h'_\varepsilon) = d(l_\varepsilon, l'_\varepsilon) + |c_\varepsilon + \varepsilon - c'_\varepsilon| \le d(l_\varepsilon, l'_\varepsilon) + \varepsilon + |f(x_\varepsilon) - f(x'_\varepsilon)|$$
$$+|l_\varepsilon(x_\varepsilon) - l'_\varepsilon(x'_\varepsilon)| \le 2s(\varepsilon) + \varepsilon + |l_\varepsilon(x_\varepsilon) - l'_\varepsilon(x'_\varepsilon)|.$$

The function ω is uniformly continuous in a δ-neighborhood of the point (x, l), so there exists $\eta(\varepsilon) \to +0$ as $\varepsilon \to +0$ and

$$|l_\varepsilon(x_\varepsilon) - l'_\varepsilon(x'_\varepsilon)| = |\omega(l_\varepsilon, x_\varepsilon) - \omega(l'_\varepsilon, x'_\varepsilon)| \le \eta(\varepsilon)$$

Thus

$$d(h_\varepsilon, h'_\varepsilon) \le 2s(\varepsilon) + \varepsilon + \eta(\varepsilon) \tag{5.6}$$

Assume now that $2s(\varepsilon) + \varepsilon + \eta(\varepsilon) \le \delta$. Then

$$d(h, h'_\varepsilon) \le d(h, h_\varepsilon) + d(h_\varepsilon, h'_\varepsilon) \le \delta + \delta = 2\delta. \tag{5.7}$$

It follows from (5.3) - 5.5) and (5.7)) that $(x'_\varepsilon, f(x'_\varepsilon), h'_\varepsilon) \in \tilde{\partial}^*_H f(2\delta; x, h)$ for sufficiently small $\varepsilon > 0$. Applying (5.4) and (5.6) we conclude that $r_{f,2\delta;x,h} \le 2s(\varepsilon) + \varepsilon + \eta(\varepsilon) \to +0$ as $\varepsilon \to +0$. □

Following [6], we say that H is $\omega - sufficient$ for X ([6]) if for each $(x, h) \in X \times H$ there exists $\delta > 0$ and a function $r_{x,h}(\varepsilon)$ such that $r_{x,h}(\varepsilon) \to 0$ as $\varepsilon \to +0$ and $r_{f,\delta;x,h} \le r_{x,h}$ for each $f \in P(H, X)$.

It was proved in [6] that if H is ω-sufficient and $f_n \to_\omega f$ then $\tilde{\partial}_H f \subset \operatorname{Li} \tilde{\partial}_H f_n$. Since ω-convergence implies H-convergence, ω-sufficiency allows us to assert that H-convergence of a sequence of H-convex functions f_n to f implies that $\tilde{\partial}_H f \subset \operatorname{Li} \tilde{\partial}_H f_n$. However, combining Theorem 3.1 with Propositions 4.2 and 5.2., we get the same conclusion under rather weaker assumption:

Theorem 5.1. *Let $f_n \in P(H, X)$ and (f_n) H-converges to $f \in PH(X)$. Suppose that the coupling function $\omega(l, x)$ is locally uniformly continuous and*

$$s_{f_n,\delta;x,h}(\varepsilon) \to 0 \quad as \ n \to +\infty \ \ and \ \varepsilon \to +0.$$

Then (f_n) ω-converges and strongly H-converges to f and

$$\tilde{\partial}_H f \subset \operatorname{Li} \tilde{\partial}_H f_n.$$

This means that verification of ω-sufficiency is a way to guarantee convergence of the graphs of subdifferential maps under H-convergence of functions. Example 3.1 describes a simple situation when all assumptions of the theorem are satisfied if we take L to be a collection of functions $l(x) = \alpha d(x, y)$. Here is another a bit more sophisticated.

Example 5.1. Let X be a Hilbert space with the inner product $[\cdot,\cdot]$. Consider the set H of all functions $h_{a,u,c}$ with $a \leq 0$, $u \in h$ and $c \in \mathbb{R}$, where

$$h_{a,u,c} = -a\|x\|^2 + [u,x] - c.$$

We can consider H as the set of all L-affine functions, where L consists of functions $h_{a,u,0}$. Assume that H is equipped with the metric d, where

$$d(h_{a,u,c}, h_{a',u',c'}) = |a - a'| + \|u - u'\| + |c - c'|.$$

Proposition 5.3. *Let $\varepsilon \in [0,1]$. Then for each $f \in P(H,X)$, $y \in X$ and $l = h_{a,u,0} \in \partial_{L,\varepsilon} f(y)$ there exist $y_\varepsilon \in X$ and $l_\varepsilon \in \partial_L f(y_\varepsilon)$ such that*

$$d(y, y_\varepsilon) \leq s_{y,l}(\varepsilon), \quad |f(y) - f(y_\varepsilon)| \leq s_{y,l}(\varepsilon), \quad d(l, l_\varepsilon) \leq s_{y,l}(\varepsilon), \tag{5.8}$$

where

$$s_{y,l}(\varepsilon) = \max(\varepsilon^{\frac{1}{3}}, \varepsilon + \varepsilon^{\frac{1}{3}} K, \varepsilon^{\frac{1}{3}} K_1),$$

and K depends on l and y, K_1 depends on y.

Proof. Let $U(x) = f(x) - l(x)$. Then $U(y) \leq \inf_{x \in X} U(x) + \varepsilon$. By the variational principle of Borwein-Preiss [3] for any positive μ there exists $v \in X$ such that

$$|U(y) - U(v)| \leq \varepsilon, \quad \|y - v\| \leq \mu \tag{5.9}$$

and the function

$$g(x) = f(x) - l(x) + \frac{\varepsilon}{\mu^2}\|x - v\|^2$$

attains its global minimum at v. Let

$$l'(x) = l(x) - (\varepsilon/\mu^2)(\|x\|^2 - 2[x,v]). \tag{5.10}$$

Then $l' \in L$. Since $g(x) \geq g(v)$ it follows that

$$\begin{aligned} f(x) - f(v) &\geq l(x) - \frac{\varepsilon}{\mu^2}\|x - v\|^2 - l(v) \\ &= l(x) - \frac{\varepsilon}{\mu^2}(\|x\|^2 - 2[x,v]) - (l(v) - \frac{\varepsilon}{\mu^2}\|v\|^2) \\ &= l'(x) - l'(v). \end{aligned}$$

Thus $l' \in \partial_L f(v)$. We now estimate $|f(y) - f(v)|$ and $d(l,l')$. Assume that $\mu \leq 1$. Then $\|v\| \leq \|y\| + 1$, so

$$\|y\| + \|v\| \leq 2\|y\| + 1. \tag{5.11}$$

Since $l \in L$, there exist $a < 0$ and $u \in X$ such that $l(x) = -a\|x\|^2 + [u, x]$. It follows from (5.9) and (5.11) that

$$|f(y) - f(v)| \leq |U(y) - U(v)| + |l(y) - l(v)| \leq \varepsilon + a|\|y\|^2 - \|v\|^2| + |[u, y - v]|$$

$$\leq \varepsilon + \|y - v\| |a(\|y| + \|v\|) + \|u\|) \leq \varepsilon + \mu K, \tag{5.12}$$

where $K = a(2\|y\| + 1) + \|u\|$. It follows from (5.10) that

$$d(l, l') = \frac{\varepsilon}{\mu^2}(1 + 2\|v\|) \leq \frac{\varepsilon}{\mu^2} K_1, \tag{5.13}$$

where $K_1 = 1 + 2(\|y\| + 1) > 1$. Put $\mu = \varepsilon^{1/3}$, $y_\varepsilon := v$ and $l_\varepsilon = l'$. Then $\mu \leq 1$ if $\varepsilon \leq 1$ and (5.8) follows directly from (5.9), (5.12) and (5.13). □

Now Propositions 5.3 and 5.2 imply that for every $(x, h) \in X \times H$ and each $\delta > 0$ there exists a function $r_{\delta;x,h}$ such that for all sufficiently small $\varepsilon > 0$ and all $f \in P(H, X)$ we have

$$r_{f,\delta;x,h}(\varepsilon) \leq r_{\delta;x,h}(\varepsilon).$$

We observe now that H is the collection of L-affine functions where L consists of all functions $l(x) = h_{a,u,0}(x) = -a\|x\|^2 + [u, x]$. Therefore we can conclude that L is ω-sufficient for X. Furthermore $\partial_L f(x)$ can be described as the collection of all pairs $(a, u) \in \mathbb{R}_+ \times X$ such that $f(x + v) - f(x) \geq -a\|v\|^2 + [x - u, v]$ for all v. We leave for the reader the formulation of the convergence result following from Theorem 5.1 in this specific situation.

6. Application to increasing functions on positive orthant

In this section we consider a situation, which are very different from the situation, described in the previous section. Consider the cone $\mathbb{R}^n_{++}$ of all n-vectors with positive coordinates. We assume that $\mathbb{R}^n_{++}$ is equipped with the coordinate-wise order relation. Consider a coupling function $\bar{\omega}(l, x) := \langle l, x \rangle$ defined on $\mathbb{R}^n_{++} \times \mathbb{R}^n_{++}$ by $\langle l, x \rangle = \min_{i \in I} l_i x_i$, where $I = \{1, \dots, n\}$. A function

$$x \mapsto \langle l, x \rangle \tag{6.1}$$

is called a min -type function. We denote the function defined by (6.1) by the same symbol l, as the vector, generating this function. Consider the set L of all min-type functions. Note that $l_k(x) \to l(x)$ for all $x \in \mathbb{R}^n_{++}$

if and only if $l_k \to l$ coordinate-wise and $l_1(x) \geq l_2(x)$ for all $x \in \mathbb{R}^n_{++}$ if and only $l_1 \geq l_2$ coordinate-wise, so we can identify L_n and $\mathbb{R}^n_{++}$ as topological ordered spaces. We also use the following notation for a vector $l \in \mathbb{R}^n_{++}$ and a number $u \geq 0$:

$$\frac{u}{l} = \left\{ \frac{u}{l_1}, \dots, \frac{u}{l_n} \right\}. \tag{6.2}$$

Let

$$H = \{h : h(x) = l(x) - c, \ l \in L, c \in \mathbb{R}\}. \tag{6.3}$$

We shall consider the pair $(\mathbb{R}^n_{++}, H)$ with the coupling function $\omega(x, (l, c)) = \langle l, x \rangle - c$.

A function $f : \mathbb{R}^n_{++} \to \mathbb{R}_{+\infty}$ is called convex-along-rays (briefly CAR) if its restriction to the ray $R_x = \{\alpha x : \alpha > 0\}$ is a convex function for each $x \in \mathbb{R}^n_{++}$. A function $f : \mathbb{R}^n_{++} \to \mathbb{R}_{+\infty}$ is called increasing if $x \leq y$ implies $f(x) \leq f(y)$. We shall use the abbreviation ICAR for increasing convex-along-rays functions. It is known (see [12]) that an ICAR function f is continuous at each point $x \in \operatorname{int}\operatorname{dom} f$. Let f be an ICAR function. Since the function $f_x(\alpha) = f(\alpha x)$ is convex, it follows that there exists directional derivatives $f'(x, x)$ and $f'(x, -x)$ which coincide with the right and left derivatives of f_x at the point $\alpha = 1$, respectively. The class of ICAR function is very broad. It contains, for example, all polynomials with nonnegative coefficients and convex increasing functions. Increasing *concave* positively homogeneous of degree one functions are ICAR as well. We shall study mainly strictly increasing CAR functions. (A function f defined on $\mathbb{R}^n_{++}$ is called strictly increasing at a point x if $(y \leq x, y \neq x \implies f(y) < f(x)$.) We need the following results, which can be found in [12], Section 3.3.

Theorem 6.1. *1) A function f defined on $\mathbb{R}^n_{++}$ is H-convex if and only if f is lower semicontinuous and ICAR;*

2) Let f be an ICAR function and $x \in \operatorname{int}\operatorname{dom} f$. Then $\partial_L f(x)$ is not empty and $\partial_L f(x) \supset A(x)$, where

$$A(x) = \left\{ l : l = \frac{u}{x} : f'(x, -x) \leq u \leq f'(x, x) \right\}. \tag{6.4}$$

3) If f is strictly increasing then $\partial_L f(x) = A(x)$.

Remark 6.1. The subdifferential $\partial_L f(x)$ can be much larger than $A(x)$ for non strictly increasing CAR functions.

6.1 IPH-functions (An example of non-convergence of support jets)

Our aim now is to give an example of a sequence f_k which H-converges to f and such the inclusion $\tilde{\partial}_H^* f \subset \mathrm{Li}\, \tilde{\partial}_H^* f_k$ does not hold. This in particular means that H-convergence does not imply strong H-convergence and the set H is not ω-sufficient for the set X. We shall consider positively homogeneous of degree one increasing (briefly IPH) functions, defined on IR^n_{++}. Clearly a function p is IPH if and only if this function ICAR and positively homogeneous of degree one. We need the following result (see [12], Subsection 2.3.7 and Subsection 7.2.7).

Theorem 6.2. *Let p be an IPH function defined on IR^n_{++}. Then*

1) $\mathrm{supp}\,(p,H) = \left\{(l,c) \in \mathrm{IR}^n_{++} \times \mathrm{IR} : p\left(\frac{1}{l}\right) \geq 1,\ c \geq 0\right\}.$
2) $\partial_L p(y) = \left\{l \in \mathrm{IR}^n_{++} : p\left(\frac{1}{l}\right) = 1,\ \langle l, y\rangle = p(y)\right\} \qquad (y \in \mathrm{IR}^n_{++});$
3) $l \in \partial_{L,\varepsilon} p(y) \Longleftrightarrow p\left(\frac{1}{l}\right) \geq 1,\ \langle l, y\rangle \geq p(y) - \varepsilon\ (y \in \mathrm{IR}^n_{++}).$

It follows from Theorem 6.2 (1) and Proposition 4.1 that the conjugate function p^ω coincides with the indicator function $\delta_{\mathrm{supp}\,(p,L)}$ of the support set:

$$p^\omega(l) = \begin{cases} 0 & p(1/l) \geq 1 \\ +\infty & p(1/l) < 1. \end{cases}$$

It is well known that $p(x) = \max\{l(x) : l \in \partial_L f(x)\}$ for an IPH function p and $x \in \mathrm{IR}^n_{++}$. So Theorem 6.2 implies also the following:

$$\partial_H^* p(x) = \partial_L^* p(x) \times \{0\}. \tag{6.5}$$

We now describe H-convergence for IPH functions (see[12], Section 7.2 for details).

Proposition 6.1. *Let p_k be a sequence of IPH functions. Then p_k H-converges to a function p if and only if p_k pointwise converges to p.*

Proof. 1) Let $p_k \xrightarrow{H} p$. Then for each x there exists a sequence $x_k \to x$ such that $p(x_k) \to p(x)$. Let $\varepsilon > 0$. Then for sufficiently large k we have $(1-\varepsilon)x \leq x_k \leq (1+\varepsilon)x$ so

$$(1/(1+\varepsilon))p_k(x_k) \leq p_k(x) \leq (1/(1-\varepsilon))p_k(x_k).$$

Thus $p_k(x) \to p(x)$.
2) Let $p_k(x) \to p(x)$ for all $x \in \mathrm{IR}^n_{++}$. Let $x \in \mathrm{IR}^n_{++}$ and $h \in \mathrm{supp}\,(p,H)$, that is (see Theorem 6.2(1)), $h = (l,c)$, where $p(1/l) \geq 1$ and $c \geq 0$. We need to find sequences $(x_k), (h_k) = (l_k, c_k)$ such that $x_k \to x$, $p_k(x_k) \to p(x)$, $l_k \to l$, $c_k \to c$ and $(l_k, c_k) \in \mathrm{supp}\,(p_k, H)$, that is, $p_k(1/l_k) \geq 1$ and

$c_k \geq 0$. We choose $x_k = x$ for all x. Then $p_k(x_k) \to p(x)$. Then we choose $c_k = c$ for all k. Turn to the choice of l_k. First assume that $p(1/l) > 1$. Then $p_k(1/l) > 1$ for sufficiently large k, so we can take $l_k = l$. Assume now that $p(1/l) = 1$. Let $\lambda_k = 1 - p_k(l)$. Take $l_k = (1/(1+\lambda_k))l$. Then $l_k \to l$ and

$$p_k\left(\frac{1}{l_k}\right) = p_k\left(\frac{1+\lambda_k}{l}\right) = 1,$$

so $(l_k, c_k) \in \operatorname{supp}(p_k, H)$. □

Example 6.1. Let $n = 2$ and $p_k(u,v) = (u^k + v^k)^{1/k}$, $k = 1, 2, \dots ,$. We have $p_k(u,v) \to p(u,v)$ for all $(u,v) \in \mathbb{R}^2_{++}$, where $p(u,v) = \max(u,v)$, so $p_k \xrightarrow{H} p$ (due to Proposition 6.1). The function p_k is strictly increasing. Calculating for this function the set A defined by (6.4) and applying Theorem 6.1, we conclude that

$$\partial_L p_k(u,v) = \left(\frac{p_k(u,v)}{u}, \frac{p_k(u,v)}{v}\right). \tag{6.6}$$

Let $(u,v) = (1,1)$. The description of the L-subdifferential $\partial_L p(1,1)$ at the point $(1,1)$ is known [12] (it easily follows from Theorem 6.2):

$$\partial_L p(1,1) = \{(s,1) : s \geq 1\} \cup \{(1,t) : t \geq 1\}.$$

Assume that

$$\tilde{\partial}^*_H p \subset \operatorname{Li} \tilde{\partial}^*_H p_k \tag{6.7}$$

Consider the triplet $z = ((1,1), 1, (s,1,0))$, where $s > 1$. It follows from (6.5) and (6.6) that $z \in \tilde{\partial}^*_H p$. Applying (6.7), we conclude that there exists a sequence $(u_k, v_k) \in \mathbb{R}^2_{++}$ such that

$$u_k \to 1, v_k \to 1, \frac{p_k(u_k,v_k)}{u_k} \to 1, \frac{p_k(u_k,v_k)}{v_k} \to s,$$

which is impossible. Thus (6.7) does not hold.

6.2 ICAR functions

Let f be an ICAR function. For each point y consider the function f_y defined on $(0,+\infty)$ by $f_y(\alpha) = f(\alpha y)$. Consider the ε-subdifferential $\partial_\varepsilon f_y(1) = [s_y^-(\varepsilon), s_y^+(\varepsilon)]$ of f_y at the point $\alpha = 1$. It is well-known that $\bigcap_{\varepsilon>0} \partial_\varepsilon f_y(1) = \partial f_y(1) = [f'(y,-y), f'(y,y)]$, so $s_y^+(\varepsilon) \to f'(y,y)$ as $\varepsilon \to +0$ and $s_y^-(\varepsilon) \to f'(y,-y)$ as $\varepsilon \to +0$.

Let f be a strictly increasing CAR function and the function f_y is differentiable at the point $\alpha = 1$. Then $\partial_L f(y)$ consists of the unique element $f'(y,y)/y$ and

$$\lim_{\varepsilon \to +0} s_y^+(\varepsilon) = \lim_{\varepsilon \to +0} s_y^-(\varepsilon) = f'(y, y).$$

Let Δ be a set of functions $\delta : \mathbb{R}_+ \to \mathbb{R}_+$ such that $\lim_{\varepsilon \to 0} \delta(\varepsilon) = 0$. Consider the set Γ_Δ of all real-valued ICAR functions f with the following properties: for each $y \in \mathbb{R}^n_{++}$ there exists $\delta_y \in \Delta$ such that
1) f is strictly increasing at the point y with the modulus δ_y, that is,

$$(x \in \mathbb{R}^n_{++}, y \geq x, \ \exists j : y_j - \delta_y(\varepsilon) \geq x_j) \Longrightarrow f(y) - \varepsilon > f(x).$$

2) The function $f_y(\alpha) = f(\alpha y)$ is differentiable at the point $\alpha = 1$.
3) $s_y^+(\varepsilon) - f'(y, y) \to 0$ and $f'(y, y) - s_y^-(\varepsilon) \to 0$ as $\varepsilon \to +0$ uniformly in a neighborhood of every point $y \in \mathbb{R}^n_{++}$.

Lemma 6.1. *Let f be a CAR function strictly increasing at a point y with the modulus δ_y, $l \in \mathbb{R}^n_{++}$, $\varepsilon > 0$ and*

$$\langle l, x \rangle - \langle l, y \rangle \leq f(x) - f(y) + \varepsilon \ \textit{for all} \ x \in \mathbb{R}^n_{++}.$$

Let $f_y(\alpha) = f(\alpha y)$. Then

$$\frac{s_y^-(\varepsilon)}{y} \leq l \leq \frac{s_y^+(\varepsilon)}{y} + \delta_y(\varepsilon)\frac{l}{y}. \tag{6.8}$$

Proof. Let $u = \langle l, y \rangle$ and $x = \alpha y$ with $\alpha > 0$. Then

$$\begin{aligned} (\alpha - 1)u &= \langle l, \alpha y \rangle - \langle l, y \rangle = \langle l, x \rangle - \langle l, y \rangle \leq f(x) - f(y) + \varepsilon \\ &= f_y(\alpha) - f_y(1) + \varepsilon, \quad \alpha > 0. \end{aligned}$$

Thus $u = \langle l, y \rangle \in \partial_\varepsilon f_y(1)$. We now prove estimates (6.8). If there exists $j \in I$ such that $l_j < u/y_j$ then $\langle l, y \rangle \leq l_j y_j < u$, which is impossible. Hence

$$l \geq \frac{u}{y} \geq \frac{s_y^-(\varepsilon)}{y}.$$

Assume there exists an index $j \in I = \{1, \dots, n\}$ such that

$$l_j > \frac{s_y^+(\varepsilon)}{y_j} + \delta_y(\varepsilon)\frac{l_j}{y_j} \geq \frac{u}{y_j} + \delta_y(\varepsilon)\frac{l_j}{y_j}.$$

Then $l_j y_j > u + \delta_y(\varepsilon) l_j$. Consider the vector z with the coordinates

$$z_i = \begin{cases} y_i & i \neq j \\ y_j - \delta_y(\varepsilon) & i = j \end{cases}$$

We have

$$\langle l, z\rangle = \min(\min_{i\neq j} l_i z_i, l_j(y_j - \delta_y(\varepsilon)) = \min(u, l_j y_j - l_j\delta_y(\varepsilon)) = u.$$

Keeping in mind that $u = \langle l, y\rangle$ (by definition), we conclude that

$$0 = \langle l, z\rangle - \langle l, y\rangle \le f(z) - f(y) + \varepsilon.$$

On the other hand, since $f \in \Gamma_\Delta$ it follows that $f(z) - f(y) + \varepsilon < 0$. Thus the result follows. □

We need the following modification of the notion of ω-sufficiency. Let H be a metric space of functions defined on a metric space X and $\Gamma \subset P(H, X)$. The set H is called ω-sufficient for X with respect to Γ if for each $(x, h) \in X \times H$ there exists $\delta > 0$ and a function $r_{x,h}(\varepsilon)$ such that $r_{x,h}(\varepsilon) \to 0$ as $\varepsilon \to +0$ and $r_{f,\delta;x,h} \le r_{x,h}$ for each $f \in \Gamma$. If H is ω-sufficient with respect to Γ then H-convergence coincides with the strong H-convergence on the set Γ.

Proposition 6.2. *Let Δ be a family of functions $\delta : \mathbb{R}_+ \to \mathbb{R}_+$ such that $\lim_{\varepsilon\to 0}\delta(\varepsilon) = 0$. Then the set H defined by (6.3) is ω-sufficient with respect to the class Γ^Δ.*

Proof. For $(y, l) \in \mathbb{R}^n_{++} \times \mathbb{R}^n_{++}$ put

$$r(\varepsilon, y, l) = \max(s^+_y(\varepsilon) - f'(y, y), f'(y, y) - s^-_y(\varepsilon), \delta_y(\varepsilon)\frac{l}{y}),$$

where $[s^-_y(\varepsilon), s^+_y(\varepsilon)] = \partial_\varepsilon f_y(1)$. It follows from the definition of the class Γ_Δ and from Lemma 6.1 that $r(\varepsilon, y, l) \to 0$ as $\varepsilon \to 0$ uniformly in a neighborhood of every point (y, l). Let $\hat{l} = f'(y, y)/y$. Then $d(\hat{l}, l) \le r(\varepsilon, y, l)$, so the pair $(\hat{y}, \hat{l})$ with $\hat{y} = y$ satisfies the definition of ω-sufficiency. □

Let $f_k \in \Gamma_\Delta$, $k = 1, 2, \ldots$ and f_k H- converges to $f \in \Gamma_\Delta$. Applying Theorem 5.1 and Proposition 6.2 we conclude that for each $(x, f(x), l) \in \tilde{\partial}_L f$ there exists a sequence $(x_k, f(x_k), l_k) \in \tilde{\partial}_L f_k$ such that $x_k \to x$, $f_k(x_k) \to f(x)$ and $l_k \to l$. Note that $l_k = f'_k(x_k, x_k)/x_k$, $l = f'(x, x)/x$. So the following result holds.

Theorem 6.3. *Let $f_k \in \Gamma_\Delta, k = 1, 2, \ldots, f \in \Gamma_\Delta$ and $f_k \xrightarrow{H} f$. Then for each $x \in \mathbb{R}^n_{++}$ there exists a sequence $x_k \in \mathbb{R}^n_{++}$ such that $x_k \to x$, $f_k(x_k) \to f(x)$ and $f'_k(x_k, x_k) \to f'(x, x)$.*

7. Some economic interpretations

Production functions, which play an important role in economic theory, often appear as the result of the following construction. Consider a producer described by a production set $Z \subset X \times Y$, where X is the set of inputs and Y is the set of outputs. Let p be a "price function", that is the function, which determines the revenue (or profit) that can be obtained from the given output y. (In the classical theory p is usually assumed linear: $p(y) = [p', y]$, where p' is a price vector.) The production function f_p, corresponding to p describes the biggest revenue, which can be obtain from the given input. In other words, $f_p(x) = \sup_{(x,y)\in Z} p(y)$. Production sets studied in modern producer theory, and all the more, production functions, sometimes have a fairly complicated structure (see e.g. [2]) which makes approximation by simpler functions necessary in certain situations.

A cost function (another important object in the economic theory) $g_p : \mathbb{R}^n_+ \to \mathbb{R}_+$ is defined by

$$g_p(y) = \inf_{(x,y)\in Z} p(x) \tag{7.1}$$

where p is a price function. The number $g_p(y)$ is the least cost of an input x that can be used for manufacturing of y.

Assume that the production set Z enjoys the following properties:

1) $(x_1, \nu y) \in Z, (x_2, \mu y) \in Z \Longrightarrow (\alpha x_1 + (1-\alpha)x_2, (\alpha\nu + (1-\alpha)\mu)y) \in Z$ for all $\alpha \in [0,1]$;
2) if $(x, y) \in Z$ and $y' \leq y$ then there exists $x' \leq x$ such that $(x', y') \in Z$

and p is a convex increasing function. These (actually weak) properties are quite natural from the economical point of view. If they hold, then the function g_p defined by (7.1) is an ICAR function.

We have already mentioned that ICAR functions may have a complicated structure. Among convenient tools for approximation ICAR functions, we would mention first of all polynomials with nonnegative coefficients (which are obviously also ICAR). To have a reasonably good approximation, we should care about approximation not only the function itself but, whenever possible, also of its derivatives. For an ICAR production and cost functions the properties 2) and 3) of the definition of Γ_Δ seem natural in which case Theorem 6.3 shows that approximation of a production function by an H-converging sequence of of differentiable production functions may be successfully used.

Fenchel-Moreau conjugacy has numerous economic interpretations. Consider, for instance, a producer, described by a production set $Z \subset X \times Y$, where X is the set of inputs and Y is the set of outputs. A pair

$(x, y) \in Z$ can be viewed as a feasible technological process with the input x and the output y. We assume that the length of a technological process is one time unit, so we consider the input x at the moment t and the output y at the moment $t+1$. This point is not very popular in static models, but it is common in dynamic models, in particular, discrete models of Ramsey and von Neumann type (see, for example, [9, 10]). Correspondingly we have two price functions, q and p, where q are prices at the moment t and p are prices at the moment $t+1$. We assume that q is fixed.

Let $g : \mathbb{R}^n_+ \to \mathbb{R}_+$ be the cost function at the moment t, corresponding to (not necessarily linear prices) $q(x)$, that is,

$$g(y) = \inf_{(x,y)\in Z} q(x). \tag{7.2}$$

Since $g(y)$ is the minimal cost of an input x, which can be used for manufacturing of y, we can consider the number $\pi(y) = p(y) - g(y)$ as the profit, which the producer can get from the production of the vector y. The number

$$\pi = \sup_y (p(y) - g(y)) \tag{7.3}$$

can be considered as the maximal profit, which the producer can get, having prices p. Assume that a certain set P of price functions is given. We can consider $(\mathbb{R}^n_+, P)$ as the pair of sets, which is coupled by the function ω: $\omega(p, y) = p(y)$. In this scheme the number π defined by (7.3) is the value of the Fenchel-Moreau (abstract) conjugate g^ω at an element $p \in P$.

We now consider a special approach to evaluation of productive activity and its profit. This approach has been systematically used by L. V. Kantorovich [8]). Having an outcome y, we are interested in a number of certain "complete" sets, which can be made from y. Assume that each complete set requires $m_i > 0$ units of the good i, $i \in I$, where $I = \{1, \dots, n\}$. Then the number y_m of complete sets, which can be produced from y is equal to $\min_i y_i/m_i$. (Note that the function $L(y) = \min y_i/m_i$ is the Leontiev production function.) The vector $m = (m_i)_{i\in I}$ can be considered as the indicator of proportions, which are required for preparation of complete sets of a given kind. Let c be the cost of a complete set. Then the price function p has the form

$$p(y) = \min_i \frac{cy_i}{m_i}. \tag{7.4}$$

Let g be the function defined by (7.2). Then the number

$$g^{\omega}(p) = \sup_{y}(\langle l, y \rangle - g(y)) \tag{7.5}$$

with $l_i = c/m_i$ and p defined by (7.4) can be considered as a the maximal profit, which can get the producer, having prices c and the indicator of proportions m. The function g^{ω} is the Fenchel-Moreau conjugate to g with respect to the set L of min-type functions. In particular, the function g^{ω} defined by (7.5) is ICAR.

A notion of the subgradient also have a very natural economic interpretation, namely a price function $p \in P$ is a P-subgradient of g at a point $y_0 \in \mathbb{R}^n_+$ if $p(y_0) - g(y_0) \geq p(y) - g(y)$. It follows from (7.3) that $\partial g(y_0)$ consists of price functions, which guarantee the biggest possible profit at the given point y_0.

Application of our results to economic models therefore requires a deeper analysis of subdifferentability and convergence properties of functions belonging to the special classes mentioned above. Today's nonsmooth analysis allows to easily get answers to some questions (e.g. relating to calculation of subdifferentials). But some questions (in particular relating to characterization of ω-convergence of some important classes of functions) remain open.

References

1. Attouch, H., Beer, G.: On the convergence of subdifferentials of convex functions. Arch. Math. **60**, 389-400 (1993)
2. Bonnisseau, J.M.: Existence of equilibria in economies with externalities and nonconvexities. Set-valued Analysis **5**, 209-226 (1997)
3. Borwein, J.M., Preiss, D.: A smooth variational principle with application to subdifferentiability and to differentiability of convex functions. Trans. Amer. Math. Soc. **303**, 517-527 (1985)
4. Dal Maso, G.: Introduction to Γ-convergence. Birkhäuser 1993
5. Intriligator, M.D.: Mathematical Optimization and Economic Theory. Prentice-Hall, Englewood Cliffs, N.J. 1971
6. Ioffe, A.D.: Abstract convexity and non-smooth analysis. Adv. Math. Economics **3**, 45-61 (2001)
7. Kutateladze, S.S., Rubinov, A.M.: Minkowski duality and its applications. Russian Mathem. Surveys **27**, 137-191 (1971)
8. Kantorovich, L.V.: Economic Calculation of the Best Use of Resources. Harvard University Press, Cambidge 1965
9. Makarov, V.L., Rubinov, A.M.: Mathematical Theory of Economic Dynamics and Equilibria. Nauka, Moskow 1973 (Russian); Springer-Verlag 1977 (English)
10. Makarov, V.L., Levin, M.J., Rubinov, A.M.: Mathematical Economic Theory: Pure and Mixed Types of Economic Mechanisms, Advanced Textbook in Economics, v.33. North-Holland Publishing Co., Amsterdam - New York 1995

11. Pallaschke, D., Rolewicz, S.: Foundations of Mathematical Optimization (Convex analysis without linearity). Kluwer Academic Publishers, Dordrecht 1997
12. Rubinov, A.M.: Abstract convexity and global optimization. Kluwer Academic Publishers 2000
13. Singer, I.: Abstract Convex Analysis. Wiley-Interscience Publication, New York 1997

Adv. Math. Econ. 4, 25–39 (2002)

Advances in
MATHEMATICAL
ECONOMICS

Dynamic programming with upper semi-continuous stochastic aggregator*

Hiroyuki Ozaki

Faculty of Economics, Tohoku University, Kawauchi, Aoba-ku, Sendai 980-8576, Japan
(e-mail: ozaki@econ.tohoku.ac.jp)

Received: May 3, 2001
Revised: July 4, 2001

JEL classification: C61, D81

Mathematics Subject Classification (2000): 90C39, 93E20

Abstract. Ozaki and Streufert (1996) develop dynamic programming techniques which are suitably adapted to the recursive objective function which is not necessarily time-separable nor state-separable. To derive their results, they assume that the stochastic aggregator, which generalizes the expectation operator, need be upper quasi-continuous. This paper proposes a new assumption on the stochastic aggregator, which we call upper *semi* -continuity. This is strictly weaker than the upper quasi-continuity. As a main result of this paper, we show that in an important class of problems called the negative dynamic programming, Ozaki and Streufert's result still holds even if we replace the assumption of upper quasi-continuity by the weaker assumption of upper semi-continuity.

Key words: Stochastic aggregator, upper semi-continuity, negative dynamic programming

1. Introduction

It is quite often to assume in economic models that the agent chooses a stochastic process $\boldsymbol{c} = (c_0, c_1, c_2, \dots)$ to maximize the objective function defined by

$$U_s(\boldsymbol{c}) = E_s \textstyle\sum_{t=0}^{\infty} \beta^t u(c_t) \tag{1}$$

* I am grateful to the anonymous referees for their helpful comments.

under some constraints. Here, s is the current state which may affect transition probabilities, E is the expectation operator conditional on the current state s, β is a discount (or upcount) factor and u is a felicity function in each period. The objective function (1) satisfies a recurrence relation:

$$\begin{aligned} U_s(\boldsymbol{c}) &= u(c_0) + \beta E_s \textstyle\sum_{t=0}^{\infty} \beta^t u(c_{t+1}) \\ &= u(c_0) + \beta E_s \left[E_{s_1} \textstyle\sum_{t=0}^{\infty} \beta^t u(c_{t+1})\right] \\ &= u(c_0) + \beta E_s U({}_1\boldsymbol{c}) \end{aligned}$$

where $U({}_1\boldsymbol{c}) = \langle U_{s_1}({}_1\boldsymbol{c}|s_1)\rangle$ is a scalar-valued function of tomorrow's state s_1. Epstein and Zin (1989), Farmer (1990) and Weil (1990) among many others motivate in economic context[1] the objective function U which satisfies a generalization of this recurrence relation:

$$U_s(\boldsymbol{c}) = W(c_0, M_s U({}_1\boldsymbol{c})) \tag{2}$$

where in economic interpretation, the dynamic aggregator W derives current utility $U_s(\boldsymbol{c})$ from current consumption c_0 and future 'mean' utility $M_s U({}_1\boldsymbol{c})$, and the stochastic aggregator M derives future 'mean' utility from the current state s and from tomorrow's random utility $U({}_1\boldsymbol{c})$. Obviously, (1) is recovered when $W(c,m) = u(c) + \beta m$ and $M_s u = E_s u$, that is, the stochastic aggregator W generalizes the time-separable structure in (1) and the stochastic aggregator M generalizes the state-separable structure in (1).

Ozaki and Streufert (1996), which is henceforth referred to as OS, develop dynamic programming techniques which are suitably adapted to the objective function satisfying (2). They provide assumptions on the model's primitives U, W and M which are sufficient for the application of these dynamic programming techniques. In particular, they develop an assumption on the stochastic aggregator M, which they call the *upper quasi-continuity.* When M is specified by the expectation operator, the upper quasi-continuity corresponds to the assumption that the sequence of probability measures converges *strongly* as the sequence of current states converges. This is, however, a strong assumption: Let $\langle s_n \rangle$ be a sequence in some metric space which converges to s_0 and let δ_s be the

[1] With their specification, the discount (upcount) factor can be varying over time, the timing of the resolution of uncertainty is relevant, and the parameter measuring the degree of risk aversion and the one measuring the degree of intertemporal substitution are disentangled. All these aspects are missing in the time-separable expected utility function (1).

point mass (or Dirac measure) at s. Then, $\langle \delta_{s_n} \rangle$ does *not* converges to δ_{s_0} strongly, although it *does* converge to δ_{s_0} weakly.

This paper proposes a new assumption on the stochastic aggregator M, which we call *upper semi-continuity*. The upper semi-continuity corresponds to the assumption that the sequence of probability measures converges *weakly* as the sequence of current states converges when M is specified by the expectation operator, and it is strictly weaker than the assumption of upper quasi-continuity employed in OS. Theorem 2 in Section 5 shows that Theorem A of OS, which guarantees the existence of the value function and characterizes it as the unique solution to Bellman's equation, still holds when we weaken the assumption of upper *quasi*-continuity to the assumption of upper *semi*-continuity provided that we add one technical assumption. In an important class of problems called the negative dynamic programming, this additional assumption follows from other assumptions, and hence, it turns out that the assumptions in this paper as a whole are strictly weaker than those assumed in OS.

In the text, we label assumptions according to OS and indicate a new assumption by putting a superscript $^-$ or $^+$, like U8$^-$.

2. Preliminaries

Let S be a Polish space (*i.e.*, a topological space which is homeomorphic to a complete separable metric space). We regard an element $s \in S$ as a single-period state which evolves exogenously. Let A be a finite set. We regard A as a set of assets and an element $x \in \mathbb{R}_+^A$ as a portfolio of investments in A. Given a topological space X, we denote the Borel σ-algebra on X by $\mathcal{B}_X$. We also denote the space of Borel-measurable functions from X into another topological space Z by $L(X, Z)$ and the space of bounded continuous functions from X into $\mathbb{R}$ by $C(X)$. A function $f : X \to \bar{\mathbb{R}} \equiv \mathbb{R} \cup \{-\infty, \infty\}$ is *upper semi-continuous* (u.s.c.) if $(\forall c \in \mathbb{R})\ \{\, x \in X \mid f(x) \geq c \,\}$ is closed. Let $\langle \mu_n \rangle_{n=1}^{\infty}$ and μ_0 be probability measures on $(S, \mathcal{B}_S)$. Then, $\langle \mu_n \rangle$ converges to μ_0 *weakly* if

$$(\forall f \in C(S)) \qquad \int_S f(s)\, \mu_n(ds) \to \int_S f(s)\, \mu_0(ds)\,,$$

while it converges to μ_0 *strongly* if

$$\sup_{E \in \mathcal{B}_S} |\mu_n(E) - \mu_0(E)| \to 0 \quad \text{as} \quad n \to \infty\,.$$

The strong convergence is equivalent to the convergence in the total variation norm and if $\langle \mu_n \rangle$ converges to μ_0 strongly, it also converges to μ_0 weakly (Stokey and Lucas, 1989). As a convention, which is not necessarily usual, we take

$$\infty - \infty = -\infty + \infty = -\infty\,. \tag{3}$$

3. Upper Semi-continuity of Stochastic Aggregator

Let $T \subseteq \bar{\mathbb{R}}$. A *stochastic aggregator* $M : S \times L(S,T) \to T$ is any function which is *monotonic*:

$$(\forall s)(\forall u' \geq u) \quad M_s u' \geq M_s u$$

where we write $M(s,u)$ as $M_s u$ for any $(s,u) \in S \times L(S,T)$. A stochastic aggregator M maps a current state and tomorrow's random utility (which is a function of tomorrow's state) into tomorrow's mean utility. We say that a stochastic aggregator M satisfies the *Fatou property* beneath a function $\bar{u} : S \to T$ if it satisfies

$$(\forall s \in S)(\forall \langle u_n \rangle_{n=1}^{\infty} \subseteq L(S,T))$$

$$(\forall n) \;\; u_n \leq \bar{u} \;\Rightarrow\; \overline{\lim_{n\to\infty}} M_s u_n \leq M_s \overline{\lim_{n\to\infty}} u_n$$

The Fatou property beneath an "integrable" function generalizes the Fatou property of the expectation operator, which is known as Fatou's lemma. Given a Polish space X, we say that a stochastic aggregator M satisfies the *extended-Feller property* on $S \times X$ beneath a function $\bar{u} : S \to T$ if it satisfies

$$(\forall \langle (s_n, x_n) \rangle_{n=1}^{\infty} \to (s_0, x_0))(\forall u \in L(S \times X, T))$$

$$u \text{ is u.s.c. and } (\forall n \geq 1)\; u(\cdot, x_n) \leq \bar{u} \;\Rightarrow\; \overline{\lim_{n\to\infty}} M_{s_n} u(\cdot, x_n) \leq M_{s_0} u(\cdot, x_0)\,.$$

The extended-Feller property extends the Feller property[2] of the expectation operator in two directions. First, it applies not only to the expectation operator but also to non-expectation operators. Second, it requires that M map an u.s.c. function which is not necessarily bounded from above to another u.s.c. function, while the Feller property requires

---

[2] If applied to M, the Feller property *would* require M to satisfy

$$(\forall f \in C(S \times X)) \quad (s,x) \mapsto M_s f(\cdot, x) \text{ is in } C(S \times X)\,.$$

See Bertsekas and Shreve (1978, p.145, Proposition 7.30) and Stokey and Lucas (1989, p.220) for the Feller property.

that M map a *bounded* continuous function to another continuous function. A stochastic aggregator M is *upper semi-continuous* (u.s.c.) on $S \times X$ beneath a function $\bar{u} : S \to T$ if it satisfies both of the Fatou property and the extended-Feller property.

OS consider the stochastic aggregator M which is upper *quasi*-continuous (u.q.c.) beneath a function $\bar{u}$:

$$(\forall \langle s_n \rangle_{n=1}^{\infty} \to s_0)(\forall \langle u_n \rangle_{n=1}^{\infty} \subseteq L(S,T))$$

$$(\forall n)\ u_n \leq \bar{u} \Rightarrow \overline{\lim_{n\to\infty}} M_{s_n} u_n \leq M_{s_0} \overline{\lim_{n\to\infty}} u_n \,.$$

The u.s.c. is a weaker requirement than the u.q.c., which is stated in the next proposition. (Actually, it turns out that the u.s.c. is *strictly* weaker than the u.q.c. See Section 3.1.)

Proposition 1. *Suppose that a stochastic aggregator M is u.q.c. beneath a function $\bar{u}$. Then for any Polish space X, M is u.s.c. on $S \times X$ beneath $\bar{u}$.*

Proof. The Fatou property immediately follows from the u.q.c. by letting $(\forall n)\ s_n = s$. To show that the extended-Feller property also follows from the u.q.c., let u be u.s.c. and such that $(\forall n)\ u(\cdot, x_n) \leq \bar{u}$. Further, define $\langle u_n \rangle$ by $(\forall n)\ u_n = u(\cdot, x_n)$. Then

$$\begin{aligned} & \overline{\lim_{n\to\infty}} M_{s_n} u(\cdot, x_n) \\ = \; & \overline{\lim_{n\to\infty}} M_{s_n} u_n \\ \leq \; & M_{s_0} \overline{\lim_{n\to\infty}} u_n \\ = \; & M_{s_0} \overline{\lim_{n\to\infty}} u(\cdot, x_n) \\ \leq \; & M_{s_0} u(\cdot, x_0) \end{aligned}$$

where the first inequality holds by the u.q.c. of M beneath $\bar{u}$, and the second inequality holds by the u.s.c. of u (Bertsekas and Shreve, 1978, p.146, Lemma 7.13(b)) and the monotonicity of M. □

3.1. An Example: Expectation Operator

A familiar example of a stochastic aggregator M is the expectation operator E^Q:

$$(\forall s \in S)(\forall u \in L(S,T)) \quad E_s^Q u = \int_S u(s')Q_s(ds') \tag{4}$$

where $Q : S \times \mathcal{B}_S \to [0,1]$ is a *stochastic kernel*, that is, Q satisfies

$$(\forall s \in S) \qquad Q_s \text{ is a probability measure on } (S, \mathcal{B}_S) \text{ and}$$

$$(\forall E \in \mathcal{B}_S) \quad Q.(E) \text{ is } \mathcal{B}_S\text{-measurable.}$$

Note that (4) is always well-defined by (3). We provide a sufficient condition for the u.s.c. of a stochastic aggregator when it is specified by the expectation operator, that is, when $M = E^Q$. To this end, the next lemma due to Gihman and Skorohod is essential.

Lemma 1 (Gihman and Skorohod, 1979, Lemma 1.5). *Let X be a Polish space, let $u \in L(S \times X, T)$ be u.s.c. and bounded from above, let $\langle \mu_n \rangle_{n=1}^{\infty}$ and μ_0 be probability measures on $(S, \mathcal{B}_S)$ such that $\langle \mu_n \rangle$ converges to μ_0 weakly, and let $\langle x_n \rangle_{n=1}^{\infty}$ be a sequence in X which converges to x_0. Then,*

$$\overline{\lim_{n\to\infty}} \int_S u(s, x_n)\, d\mu_n(s) \leq \int_S u(s, x_0)\, d\mu_0(s)\,.$$

Theorem 1 (Sufficient Condition for U.S.C.). *Let X be a Polish space, let Q be a stochastic kernel and define the stochastic aggregator M by $M = E^Q$. Then, M is u.s.c. on $S \times X$ beneath $\bar{u} \in L(S,T)$ if $(Q, \bar{u})$ satisfies:*

M1$^-$. $(\forall \langle s_n \rangle_{n=1}^{\infty} \to s_0) \quad Q_{s_n}$ converges to Q_{s_0} *weakly* as $n \to \infty$ and

M2. $(\forall \langle s_n \rangle_{n=1}^{\infty} \to s_0) \quad \displaystyle\lim_{b\to\infty} \overline{\lim_{n\to\infty}} \int_{\{ s' \mid \bar{u}(s') \geq b \}} \bar{u}(s')\, Q_{s_n}(ds') = 0\,.$

Proof. The Fatou property follows from Fatou's lemma since M2 implies that $\bar{u} \vee 0$ is integrable with respect to Q_s for any s (let $(\forall n)\ s_n = s$). To prove that the extended-Feller property also follows, let $(s_n, x_n) \to (s_0, x_0)$ and let $b \in \mathbb{R}_+$. Then

$$
\begin{aligned}
&\overline{\lim_{n\to\infty}} \int_S u(\cdot, x_n)\, dQ_{s_n} \\
\leq\ & \overline{\lim_{n\to\infty}} \left[\int_S u(\cdot, x_n)\, dQ_{s_n} - \int_S (u(\cdot, x_n) \wedge b)\, dQ_{s_n} \right] \\
&\qquad\qquad\qquad\qquad + \overline{\lim_{n\to\infty}} \int_S (u(\cdot, x_n) \wedge b)\, dQ_{s_n} \\
\leq\ & \overline{\lim_{n\to\infty}} \left[\int_S u(\cdot, x_n)\, dQ_{s_n} - \int_S (u(\cdot, x_n) \wedge b)\, dQ_{s_n} \right] + \int_S u(\cdot, x_0)\, dQ_{s_0} \\
=\ & \overline{\lim_{n\to\infty}} \int_{\{\, u(\cdot, x_n) \geq b \,\}} (u(\cdot, x_n) - b)\, dQ_{s_n} + \int_S u(\cdot, x_0)\, dQ_{s_0} \\
\leq\ & \overline{\lim_{n\to\infty}} \int_{\{\, \bar{u} \geq b \,\}} \bar{u}\, dQ_{s_n} + \int_S u(\cdot, x_0)\, dQ_{s_0}\,,
\end{aligned}
$$

where the second inequality follows from Lemma 1, M1$^-$, the fact that $u \wedge b$ is u.s.c. and bounded from above and the fact that $u \geq u \wedge b$. The third inequality follows because $(\forall n)\ u(\cdot, x_n) \leq \bar{u}$ and because $(\forall n)\ u(\cdot, x_n) - b \leq \bar{u}$ since $b \geq 0$. Because this whole inequality holds for all $b \geq 0$, M2 implies the desired inequality. □

Note that if Q is constant with respect to s, then M1$^-$ is satisfied and M2 is equivalent to the integrability of $\bar{u} \vee 0$. A stochastic kernel Q which satisfies M1$^-$ is said to be *continuous* (Bertsekas and Shreve, 1978, p.134, Definition 7.12). The expectation operator with a continuous stochastic kernel maps a bounded continuous function to another bounded continuous function (Feller property. Bertsekas and Shreve, 1978, p.145, Proposition 7.30) and it maps an u.s.c. function bounded from above to another u.s.c. function bounded from above (Lemma 1 above or Bertsekas and Shreve, 1978, p.148, Proposition 7.31(b)). The u.s.c. of a stochastic aggregator does not require the integrand to be bounded, and that's why we needed an additional assumption, M2.

OS (Lemma C.1) show that the u.q.c. is implied by M2 and

$$
\text{M1.} \quad (\forall \langle s_n \rangle_{n=1}^{\infty} \to s_0) \quad \sup_{E \in \mathcal{B}_S} |Q_{s_n}(E) - Q_{s_0}(E)| \to 0 \ \text{ as } \ n \to \infty\,.
$$

Assumption M1 is equivalent to requiring that Q_{s_n} converge to Q_{s_0} *strongly* whenever $s_n \to s_0$. Theorem 1 substantiates the claim that the u.s.c. is a weaker requirement for a stochastic aggregator M than the u.q.c. when M is specified by the expectation operator since the weak convergence is a strictly weaker requirement for a stochastic kernel Q than the strong convergence. To make this point clear, consider the stochastic kernel $\widehat{Q}$ defined by

$$(\forall s) \quad \widehat{Q}_s(\{s\}) = 1 .$$

Note that since a one-point set in a metric space is closed and hence Borel-measurable, $\widehat{Q}$ is well-defined. The stochastic kernel $\widehat{Q}$ corresponds to the deterministic process which stays forever at the initial state. Then, it follows that $\widehat{Q}$ satisfies M1$^-$, but not M1.[3]

We may use $\widehat{Q}$ also to show that the u.s.c. is a *strictly* weaker requirement for a stochastic aggregator than the u.q.c. Define $\widehat{M}$ by $\widehat{M} = E^{\widehat{Q}}$. Since M2 is clearly satisfied when $\bar{u}$ is bounded from above and since $\widehat{Q}$ is continuous, $\widehat{M}$ is u.s.c. beneath any bounded function by Theorem 1. However, $\widehat{M}$ is *not* u.q.c. beneath a bounded function: Let $\langle s_n \rangle \subseteq S$ be such that $(\forall n)\ s_n \neq s_0$ and $s_n \to s_0$, and let $\langle u_n \rangle$ be such that $(\forall n)\ u_n(s) = 0$ if $s = s_0$ and $u_n(s) = 1$ otherwise. (Recall that u.q.c. does not require the integrand to be an u.s.c. function.) Then, $(\forall n)\ u_n \leq 1$ and

$$\overline{\lim_{n\to\infty}} \widehat{M}_{s_n} u_n = \overline{\lim_{n\to\infty}} u_n(s_n) = 1 > 0 = \overline{\lim_{n\to\infty}} u_n(s_0) = \widehat{M}_{s_0} \overline{\lim_{n\to\infty}} u_n .$$

The discussion thus far suggests that an u.s.c. stochastic aggregator generalizes an expectation operator with a continuous stochastic kernel.

We state the remark made in the previous paragraph as a corollary.

Corollary 1. *Let X be a Polish space, let Q be a stochastic kernel and define the stochastic aggregator M by E^Q. Also assume that $\bar{u}$ is bounded from above (by a finite constant). Then, M is u.s.c. on $S \times X$ beneath $\bar{u}$ if Q satisfies:*

M1$^-$. $\quad (\forall \langle s_n \rangle_{n=1}^{\infty} \to s_0) \quad Q_{s_n}$ converges to Q_{s_0} *weakly* as $n \to \infty$.

[3] Let $\langle s_n \rangle \subseteq S$ be such that $(\forall n)\ s_n \neq s_0$ and $s_n \to s_0$. Then,

$$(\forall f \in C(S)) \quad \lim_{n\to\infty} \int_S f\, d\widehat{Q}_{s_n} = \lim_{n\to\infty} f(s_n) = f(s_0) = \int_S f\, d\widehat{Q}_{s_0} ,$$

while $\lim_{n\to\infty} \widehat{Q}_{s_n}(\{s_0\}) = 0 \neq 1 = \widehat{Q}_{s_0}(\{s_0\})$.

4. Dynamic Optimization Problem

We state the dynamic optimization problem with an economic interpretation according to OS. Let $F : S \times \mathbb{R}_+^A \to \mathbb{R}_+$ satisfy[4]

F1. F is weakly increasing in its second argument and u.s.c. in its both arguments.

We regard $F_s(x_-)$ as today's income, given that today's state is s and that yesterday's investment portfolio was x_-. Let $t \in \{0, 1, 2, \dots\}$ denote time. We denote a history of states' realization up to time t by $(s_1, \dots, s_t)$ or ${}_1s_t$. A *consumption process* $\boldsymbol{c} = (c_0, c_1, c_2, \dots)$ is an element of $\Pi_{t \geq 0} L(S^t, \mathbb{R}_+)$, where $L(S^0, \mathbb{R}_+) \equiv \mathbb{R}_+$. Similarly, a *portfolio process* $\boldsymbol{x} = (x_0, x_1, x_2, \dots)$ is an element of $\Pi_{t \geq 0} L(S^t, \mathbb{R}_+^A)$, where $L(S^0, \mathbb{R}_+^A) \equiv \mathbb{R}_+^A$. A consumption process $\boldsymbol{c}$ is *feasible* from an initial income $y \in \mathbb{R}_+$ if there exists a corresponding portfolio process $\boldsymbol{x}$ such that $c_0 + \Sigma_{a \in A} x_0^a \leq y$ and $(\forall t \geq 1)$ $c_t + \Sigma_{a \in A} x_t^a \leq F(x_{t-1})$. From the function F, derive the function $\bar{F} : S \times \mathbb{R}_+ \to \mathbb{R}_+$ by

$$\bar{F}_s(y_-) = \max\{\, F_s(x_-) \mid \Sigma_{a \in A} x_-^a \leq y_- \,\}\,.$$

The existence and u.s.c. of $\bar{F}$ is insured by F1 and the Maximum Theorem (Berge, 1962, p.116, Theorem 2). The quantity $\bar{F}_s(y_-)$ bounds today's income as a function of yesterday's income y_- and today's state s. Let $\bar{F}^t$ denote the t-fold composition of $\bar{F}$, that is, $\bar{F}^t(y) = \bar{F}_{s_t}(\bar{F}_{s_{t-1}}(\cdots(\bar{F}_{s_1}(y))\cdots))$ (let $\bar{F}^0(y) = y$). Then,

Lemma 2. *For any $t \geq 1$, $\bar{F}^t : S^t \times \mathbb{R}_+ \to \mathbb{R}_+$ is u.s.c. and weakly increasing in y.*

Proof. The proof is by induction. The claim holds for $t = 1$ since $\bar{F}$ is u.s.c. by the remark made in the previous paragraph and since it is weakly increasing in y by its definition. Suppose that the claim holds for $t - 1$. Then, $\bar{F}^t = \bar{F}(\bar{F}^{t-1})$ is u.s.c. by $\bar{F}$'s u.s.c. and weak increase in y and by the inductive hypothesis, and it is weakly increasing in y by $\bar{F}$'s weak increase in y and the inductive hypothesis. □

It follows from Lemma A.1 of OS that every feasible consumption process from some initial income lies in

[4] We slightly strengthen Assumption F1 of OS by requiring F's weak increase. This is because we state Assumption U8$^-$ below in terms of F, rather than $\bar{F}$ whose weak increase is automatic.

$$\boldsymbol{C} \equiv \{\, \boldsymbol{c} \in \Pi_{t\geq 0} L(S^t, \mathbb{R}_+) \mid (\exists y)\ \boldsymbol{c} \leq {}_0\bar{\boldsymbol{F}}(y) \,\},$$

where ${}_0\bar{\boldsymbol{F}}(y)$ denotes the "pure-accumulation" process $\langle\, \bar{F}^t(y) \,\rangle_{t\geq 0}$.

Preferences are specified by a utility space $T \subseteq \bar{\mathbb{R}}$, a *utility function* $U : S \times \boldsymbol{C} \to T$, a *dynamic aggregator* $W : \mathbb{R}_+ \times T \to T$, and a *stochastic aggregator* $M : S \times L(S,T) \to T$. The utility space T is assumed to be closed, and U, W and M are assumed to satisfy

U1. $(\forall s)(\forall \boldsymbol{c}' \geq \boldsymbol{c}) \quad U_s(\boldsymbol{c}') \geq U_s(\boldsymbol{c})$;

U2. $(\forall s)(\forall y)(\forall \boldsymbol{c} \leq {}_0\bar{\boldsymbol{F}}(y)) \quad \lim_{t\to\infty} U_s({}_0\boldsymbol{c}_{t-1}, {}_t\bar{\boldsymbol{F}}(y)) = U_s(\boldsymbol{c})$;

U4. $(\forall s)(\forall \boldsymbol{c}) \quad U_s(\boldsymbol{c}) = W(c_0, M_s U({}_1\boldsymbol{c}))$;

U5. W is weakly increasing in its second argument;

U6. W is u.s.c.; and

U7. $(\forall s)(\forall u' \geq u) \quad M_s u' \geq M_s u$.

Assumptions U1 and U2 assume that U is monotonic and upper-convergent. In U2, ${}_0\boldsymbol{c}_{t-1}$ denotes the restriction of the process $\boldsymbol{c}$ to periods $\{\, 0, 1, \ldots, t-1 \,\}$ and ${}_t\bar{\boldsymbol{F}}(y)$ denotes the restriction of the pure-accumulation process ${}_0\bar{\boldsymbol{F}}(y)$ to periods $\{\, t, t+1, \ldots \}$. Assumption U4 is a stochastic version of Koopmans' (1960) equation. Note that $U({}_1\boldsymbol{c})$ is a scalar-valued function of tomorrow's state s_1 (it might be written more explicitly as $\langle U_{s_1}({}_1\boldsymbol{c}|s_1)\rangle_{s_1\in S}$, where ${}_1\boldsymbol{c}|s_1$ denotes the continuation of the stochastic process $\boldsymbol{c}$ at and after the state s_1). Also note that U4 requires that $U({}_1\boldsymbol{c}) \in L(S,T)$ since the domain of M is $L(S,T)$. For the intuition behind U2 and U4, the readers are referred to OS. Assumptions U5-U7 are straightforward. (We already assumed U7 by requiring M to be monotonic.)

We impose two more assumptions. The first assumption is a counterpart of U8 assumed in OS and uses the concept of u.s.c. of M:

U8$^-$. $(\forall x \in \mathbb{R}_+^A)$
M is upper *semi*-continuous on $S \times \mathbb{R}_+^A$ beneath $J^+ \circ F(x)$,

where a function $J^+ : S \times \mathbb{R}_+ \to T$ is defined by

$$(\forall s)(\forall y) \quad J_s^+(y) = U_s({}_0\bar{\boldsymbol{F}}(y)),$$

which is well-defined since $(\forall y)\ {}_0\bar{\boldsymbol{F}}(y) \in \boldsymbol{C}$ by Lemma 2 and interpreted as an extremely optimistic value function since any consumption process that is feasible from y is bounded from above by the pure-accumulation process ${}_0\bar{\boldsymbol{F}}(y)$. Note that we write $J^+_s(F_s(x))$ as $J^+ \circ F(x)$. Assumption U8 in OS assumes that $(\forall y_- \in \mathbb{R}_+)$ M is upper *quasi*-continuous beneath $J^+ \circ \bar{F}(y_-)$. Note that we use $J^+ \circ F$ rather than $J^+ \circ \bar{F}$. Apart from this slight difference, U8$^-$ substantially weakens U8 since the u.s.c. is a strictly weaker requirement than the u.q.c. as discussed in the previous section.

The second assumption is new:

$$\text{U9}^+. \quad J^+ \text{ is u.s.c.}$$

In OS (Lemma A.3), U9$^+$ was *derived*, rather than *assumed*, from U8 and other assumptions. Here, we can*not* derive U9$^+$ under U8$^-$ in a general setting of this section, and hence we need to assume it. Therefore, Assumption U9$^+$ could be seen as a cost of weakening U8 to U8$^-$. However, we will present in Section 6 an important class of dynamic optimization problems in which U9$^+$ is derived from other assumptions, and hence, in which the set of assumptions as a whole is certainly weaker than that of OS.

5. Main Result

Define the *true value function* $J^* : S \times \mathbb{R}_+ \to T$ by

$$J^*_s(y) = \max\{\, U_s(\boldsymbol{c}) \mid \boldsymbol{c} \text{ is feasible from } y \,\}\,.$$

A consumption process $\boldsymbol{c}$ is *optimal* from y and s if it is feasible from y and satisfies $U_s(\boldsymbol{c}) = J^*_s(y)$. Define the *policy correspondence* K^* : $S \times \mathbb{R}_+ \twoheadrightarrow \mathbb{R}_+ \times \mathbb{R}^A_+$ by

$$K^*_s(y) = \arg\max\{\, W(c, M_s J^* \circ F(x)) \mid c + \Sigma_{a\in A} x^a \le y \,\}\,.$$

A consumption process $\boldsymbol{c}$ is *recursively optimal* from y and s if there exists a corresponding portfolio process $\boldsymbol{x}$ such that $(c_0, x_0) \in K^*_s(y)$ and $(\forall t \ge 1)$ $(c_t, x_t) \in K^* \circ F(x_{t-1})$. A function $J : S \times \mathbb{R}_+ \to T$ is an *admissible value function* if it is u.s.c., is weakly increasing in its second argument and satisfies $J \le J^+$. Define *Bellman's operator* B from the set of admissible value functions into itself by

$$(\forall s)(\forall y) \quad BJ_s(y) = \max\{\, W(c, M_s J \circ F(x)) \mid c + \Sigma_{a\in A} x^a \le y \,\}\,.$$

Finally, we say that J *solves Bellman's equation* if $J = BJ$. The next theorem is a counterpart of Theorem A in OS.

Theorem 2 (Theorem A′). *Assume F1, U1-U2, U4-U7, U8⁻ and U9⁺. Then, (1) B is well-defined (that is, BJ exists and is admissible whenever J is admissible); (2) J^* exists, it is the greatest*[5] *admissible solution to Bellman's equation and it equals*[6] $\lim_{n\to\infty} B^n J^+$; *and (3) K^* exists, it admits a measurable selection and recursive optimality implies optimality.*

Proof. OS successively show Lemma A.2, Lemma A.3, Lemma A.5 and Proof A.6 to prove Theorem A. Hence, we need to show that the claims of these lemmas still hold when we replace U8 (OS) by U8^- plus U9^+ (here) whenever U8 is invoked, which occupies the rest of this proof.

(Lemma A.2.) In the proof of lemma, we needed to prove

$$\overline{\lim_{n\to\infty}} M_{s^n} J^\circ \circ F(x^n) \le M_{s^0} J^\circ \circ F(x^0)$$

where $s^n \to s^0$, $x^n \to x^0$ and J° is an admissible function. F1 and the admissibility of J° imply that $J^\circ \circ F$ is u.s.c. and $(\forall n)\ J^\circ \circ F(x^n) \le J^\circ \circ F(\vee_{n=1}^\infty x^n) \le J^+ \circ F(\vee_{n=1}^\infty x^n)$, where $\vee$ denotes the lattice supremum. Hence, U8^- (the extended-Feller property) immediately proves the above inequality.

(Lemma A.3.) The claim of the lemma is now assumed in U9^+.

(Lemma A.5.) There is no direct use of U8 in the proof of lemma.

(Proof A.6.) In this proof, we needed to prove

$$\overline{\lim_{m\to\infty}} M_s B^{\eta(m)} J^+ \circ F(x^{\eta(m)}) \le M_s \overline{\lim_{m\to\infty}} B^{\eta(m)} J^+ \circ F(x^{\eta(m)})$$

where $s \in S$, $\eta(m) \to \infty$ as $m \to \infty$ and $x^{\eta(m)} \to \hat{x}$ for some $\hat{x}$ as $m \to \infty$. Since B maps an admissible function to another admissible function by Lemma A.5, $(\forall m)\ B^{\eta(m)} J^+ \le J^+$. This, F1 (the weak increase of F) and the fact that J^+ is weakly increasing in y imply that $\langle B^{\eta(m)} J^+ \circ F(x^{\eta(m)}) \rangle_{m=1}^\infty$ is bounded from above by $J^+ \circ F(\vee_{m=1}^\infty x^{\eta(m)})$. Hence, U8^- (the Fatou property) proves the above inequality. □

6. Negative Dynamic Programming

This section, which is corresponding to Section 4 of OS, sets $T = \bar{\mathbb{R}}_- \equiv \mathbb{R}_- \cup \{-\infty\}$ and constructs a negative-valued (nonpositive-valued) utility function U from a dynamic aggregator W and a stochastic aggregator M. We then develop assumptions for W and M which are sufficient

[5] That is, J^* is an admissible solution to Bellman's equation and $(\forall s)(\forall y)\ J'_s(y) \le J^*_s(y)$ for any other admissible solution J'.

[6] The limit is pointwise.

to imply Assumptions U1-U2, U4-U7, U8$^-$ and U9$^+$ of Section 5's Theorem 2 and discuss that these assumptions as a whole are strictly weaker than those assumed in OS.

Suppose that a dynamic aggregator $W : \mathbb{R}_+ \times \bar{\mathbb{R}}_- \to \bar{\mathbb{R}}_-$ and a stochastic aggregator $M : S \times L(S, \bar{\mathbb{R}}_-) \to \bar{\mathbb{R}}_-$ satisfy

B1. W is weakly increasing ;

B2. W is u.s.c. ;

B3. $(\forall s)(\forall u' \geq u) \quad M_s u' \geq M_s u$;

B4. $(\forall s)(\forall y)(\forall$ weakly decreasing $\langle u_n \rangle_{n=1}^{\infty} \leq W(\bar{F}(y), 0))$
$\lim_{n\to\infty} M_s u_n = M_s \lim_{n\to\infty} u_n$;

B5. $(\forall t)(\forall u_{t+1} \in L(S^{t+1}, \bar{\mathbb{R}}_-))$
the map ${}_1s_t \mapsto M_{s_t} u_{t+1}({}_1s_t, \cdot)$ is in $L(S^t, \bar{\mathbb{R}}_-)$; and

B6$^-$. For any Polish space X, M is *u.s.c.* on $S \times X$ beneath 0 .

Assumptions B1-B5 are identical to those assumed in OS. Assumption B6$^-$ is new. We will show that this is a fairly weak assumption later (see Corollary 2 and its proof). Define $U : S \times \boldsymbol{C} \to \bar{\mathbb{R}}_-$ by

$$(\forall s)(\forall \boldsymbol{c}) \quad U_s(\boldsymbol{c}) = \lim_{t\to\infty} W(c_0, M_s W(c_1, M \cdots W(c_t, 0) \cdots)) . \tag{5}$$

The next theorem substantially strengthens Theorem E of OS.

Theorem 3 (Theorem E′). *Assume F1, B1-B5 and B6*$^-$. *Then U is well-defined and satisfies all the assumptions of Theorem 2.*

Proof. The well-definedness, U1, U2 and U4-U7 follow from Theorem E of OS. Assumption U8$^-$ follows from B6$^-$ since $(\forall x)\ J^+ \circ F(x) \leq 0$ by the fact that $T = \bar{\mathbb{R}}_-$. It remains to show that Assumption U9$^+$ is satisfied. To do this, note that

$$\begin{aligned} J_s^+(y) &= \lim_{t\to\infty} W(y, M_s W(\bar{F}(y), M \cdots W(\bar{F}^t(y), 0) \cdots)) \\ &= \inf_{t\geq 0} W(y, M_s W(\bar{F}(y), M \cdots W(\bar{F}^t(y), 0) \cdots)) , \end{aligned} \tag{6}$$

where the first equality holds by (5) and the second equality holds since the sequence defining the limit is weakly decreasing by B1, B3 and the fact that $T = \bar{\mathbb{R}}_-$.

Let $t \geq 0$. This paragraph shows by backward induction on $r \in \{0, 1, \ldots, t\}$ that the function defined by

$$({}_1s_r, y) \mapsto W(\bar{F}^r(y), MW(\bar{F}^{r+1}(y), M \cdots W(\bar{F}^t(y), 0) \cdots)) \quad (7)$$

is u.s.c. The initial step ($r = t$) holds by Lemma 2, B1 and B2. The inductive step holds because (a) the function defined by

$$({}_1s_r, y) \mapsto MW(\bar{F}^{r+1}(y), M \cdots W(\bar{F}^t(y), 0) \cdots)$$

is u.s.c. by the inductive hypothesis ($({}_1s_{r+1}, y) \mapsto W(\bar{F}^{r+1}(y), M \cdots W(\bar{F}^t(y), 0) \cdots)$ is u.s.c.), the fact that $T = \bar{\mathbb{R}}_-$ and B6$^-$ (set $X = S^r \times \mathbb{R}_+$); and (b) (7) is u.s.c. by Lemma 2, (a), B1 and B2.

Hence we may conclude that the function defined by

$$(s, y) \mapsto W(y, M_s W(\bar{F}(y), M \cdots W(\bar{F}^t(y), 0) \cdots))$$

is u.s.c. (set $r = 0$ in (7)). Since this holds for all t and since J^+ is the infimum of u.s.c. functions by (6), J^+ itself is u.s.c. (Rudin, 1987, p.38). □

Theorem E of OS assumes B1-B5 and shows that U is well-defined and satisfies U1, U2 and U4-U7. It does *not* prove that U satisfies U8. Rather, Propositions 5 and 6 of OS show that U8 is satisfied by some classes of examples. Here, in contrast, we assume B6$^-$, which is a strictly weaker assumption than U8 when adapted to negative utilities, and then show that U satisfies *all* the assumptions of Theorem 2 including U8$^-$ and U9$^+$. Hence, Theorem 3 strengthens Theorem E of OS. This fact is highlighted by the next corollary which only requires that the stochastic kernel be continuous when M is specified by the expectation operator.

Corollary 2. *Suppose that the stochastic aggregator M is given by* (4) *with a continuous stochastic kernel, that is, a stochastic kernel which satisfies M1$^-$. Also assume F1 and that the dynamic aggregator W satisfies B1 and B2. Then U is well-defined and satisfies all the assumptions of Theorem 2.*

Proof. It suffices to show that Assumptions F1, B1-B5 and B6$^-$ are all met. F1, B1 and B2 are now assumed. B3 holds obviously. B4 holds by the Lebesgue monotone convergence theorem since $(\forall y)\ W(\bar{F}(y), 0) \leq 0$ by the fact that $T = \bar{\mathbb{R}}_-$. B5 holds by Bertsekas and Shreve (1978, p.144, Proposition 7.29). Finally, B6$^-$ holds by Corollary 1. □

References

Berge, C.: Topological Spaces. Oliver and Boyd, Edinburgh 1963

Bertsekas, D.P., Shreve, S.E.: Stochastic Optimal Control: The Discrete Time Case. Academic Press, New York 1978

Epstein, L.G., Zin, S.E.: Substitution, risk aversion, and the temporal behavior of consumption and asset returns: A theoretical framework. Econometrica **57**, 937-970 (1989)

Farmer, R.E.A.: RINCE preferences. Quarterly Journal of Economics **105**, February, 43-60 (1990)

Gihman, I.I., Skorohod, A.V.: Controlled Stochastic Processes. Springer-Verlag, New York 1979

Koopmans, T.C.: Stationary ordinal utility and impatience, Econometrica **28**, 287-309 (1960)

Ozaki, H., Streufert, P.A.: Dynamic programming for non-additive stochastic objectives. Journal of Mathematical Economics **25**, 391-442 (1996)

Rudin, W.: Real and Complex Analysis (Third Edition). McGraw-Hill 1987

Stokey, N.L., Lucas, R.E.Jr.: Recursive Methods in Economic Dynamics. Harvard University Press, Cambridge, MA 1989

Weil, P.: Nonexpected utility in macroeconomics. Quarterly Journal of Economics **105**, February, 29-42 (1990)

Adv. Math. Econ. 4, 41–59 (2002)

Advances in
MATHEMATICAL
ECONOMICS

On the existence of equilibria of equicontinuous games with incomplete information

Hiroshi Tateishi*

School of Economics, Nagoya University, Furo-cho, Chikusa-ku, Nagoya, 464-8601, Japan
(E-mail:tateishi@soec.nagoya-u.ac.jp)

Received: September 4, 2001
Revised: September 28, 2001

JEL classification: C62, C72

Mathematics Subject Classification (2000): 28 A 33, 28 A 50, 91 B 52

Abstract. This paper concerns the existence of equilibria of games with incomplete information. The results obtained so far assumed some regularity condition on the prior. By restricting each player's strategy set to equicontinuous one, we establish the existence of equilibria without having recourse to the regularity condition.

Key words: Bayesian game, ϵ-approximate equilibrium, equicontinuous game, incomplete information, Young measure, transition probability.

1. Introduction

A game with incomplete information is a game in which some or all of the players lack full information about the rules of the game. Following the

* Part of this paper was written during my visit to the Department of Economics, University of California at Berkeley in May, 2001. I thank the members of the faculty for their hospitality. I am gratefully indebted to Robert Anderson, Ruca Rigotti, and Chris Shannon for their very valuable comments. I am especially grateful to Robert Anderson who made my visit at the Department possible. This paper was presented at the seminar held at the Department of Economics, Ohio State University. I gratefully acknowledge the stimulating comments of Tatsuro Ichiishi. I would like to give special thanks to an anonymous referee for detailed comments on a first draft. This work was partially supported by Nitto Foundation and Shikishima Foundation.

Bayesian approach, it is assumed that each player will assign a subjective probability distribution to all variables unknown to him in such a game. But this approach generates "an infinite hierarchy of beliefs (i.e., subjective probability distributions)" for each players since each player's belief about another player's subjective probability comes into question. This is the main difficulty in analyzing the incomplete information games. Harsanyi[11] developed a method to analyze such a situation. To overcome this obstacle, Harsanyi introduced the concept "type". A player's type specifies all the relevant private information (information that is not common knowledge) he possesses beforehand. Harsanyi represents all parameters of the game and beliefs concerning player i by one vector which he calls the "attribute vector of player i." Then, each player has a number of possible "types" corresponding to the values his attribute vector can take. Also each player will have the subjective probability distribution on other players' types while knowing his own type. This probability distribution is called the *prior* of player i.

In this way, the game with incomplete information can be transformed to the game with complete information in which the initial node is the node of the realization of the priors and this complete information game is called the *Bayesian game.* We assume moreover that the consistency assumption holds that each player's prior given his type can be derived as the conditional probability distribution of a certain basic probability. (For the rigorous mathematical construction of type sets and the prior, see Mertens-Zamir[15].)

In this paper, we consider the existence of equilibria of a Bayesian game. The existence problem has been examined by several authors (see, for example, Ambruster-Böge[1], Balder[3], Mamer-Schilling[14], and Milgrom-Weber[16]). The results obtained so far required the following regularity condition[1]: the prior is absolutely continuous with respect to the product of its marginal probability distributions[2]. When the type set is finite or the marginal probabilities are independent, this regularity condition is certainly satisfied. The following example illuminates the case where this condition does not hold.

---

[1] One exception is the work of Mamer-Schilling[14]. Remark that this condition is used to show the joint continuity of payoff functions. Mamer-Schilling[14] dealt with a special case of two person zero-sum games (see Remark 1 below). In this case, Sion's minimax theorem can be applied to establish the existence of equilibria. And so, it is sufficient to verify the continuity of payoff functions in each variable separately.

[2] Let $(T, \mathcal{T}, \mu)$ be a probability measure space with $T = T_1 \times T_2$ and p_i $(i = 1, 2)$ be the projection of $T_1 \times T_2$ on T_i, i.e., $p_i(t_1, t_2) = t_i$. Then, the marginal probability μ_i on T_i is defined by $\mu_i = \mu \circ p_i^{-1}$. And so, when μ is the prior and T_i is the type set of player i, this condition states that μ is absolutely continuous with respect to $\mu_1 \otimes \mu_2$.

Example 1. Assume that there are two players $i = 1, 2$. Let $T_i = [0,1]$ be player i's type set and $T = [0,1] \times [0,1]$. The prior μ on T is assumed to be uniformly distributed on the diagonal of the square T, i.e., on $D = \{(x, y) \in T : x = y\}$. Then, the marginal distribution μ_i on T_i is the Lebesgue measure. Hence $\mu_1 \otimes \mu_2(D) = 0$, although $\mu(D) = 1$. Accordingly, μ is not absolutely continuous with respect to $\mu_1 \otimes \mu_2$.

Furthermore, this condition does not hold in some applications. One of the representative examples is the common value auction, where the value of the object is the same to all bidders but each bidder may have different estimates of the object due to the difference of each bidder's private information. In this case, the existence of public information causes the condition difficult to meet.

The main concern of this paper is to establish the existence of equilibria without having recourse to the regularity condition. We make some restriction on each player's strategy set to resolve the problem. In the Bayesian game, a player's (mixed) strategy is defined by a (measurable) function from his type set to mixtures of his action set. We assume moreover that each player's strategies form an equicontinuous family of functions (for the precise meaning, see Section 4). We call a game whose strategy sets are equicontinuous families of functions an *equicontinuous game*. By restricting the strategy sets to equicontinuous ones, we can deduce the existence of equilibria of the Bayesian game in the case that each player's type set is a compact metric space. Notice that, since the uncertainty of the game can be regarded as the uncertainty of utility functions which is assumed to be bounded and continuous, the set of all variables of the game unknown to each player is a separable metric space. And thus, the type set is a separable metric space in general. However, in this case, we do not succeed in establishing the existence of equilibria and we are going to demonstrate the existence of ε-approximate equilibria instead.

We make use of the mathematical concept of "Young measure" to represent the strategies of the game. And the next section is devoted to explain this concept.

2. Notations and propositions

Let $(T, \mathcal{T}, \mu)$ be a complete finite positive measure space and (S, d) be a complete separable metric space. We suppose that S is endowed with the Borel σ-algebra $\mathcal{B}(S)$, i.e., the σ-algebra generated by all open subsets of S. And denote by $\mathcal{M}^1_+(S)$ the set of probability measures on S equipped with the *narrow topology*, i.e., the weakest topology which makes the linear functionals defined on $\mathcal{M}^1_+(S)$ continuous:

$$\lambda \longmapsto \int_S \psi d\lambda \quad (\psi \in \mathcal{C}^b(S)),$$

where $\mathcal{C}^b(S)$ stands for the space of continuous bounded real-valued functions defined on S.

$\mathcal{Y}(T, \mu; S)$ denotes the set of positive measures λ on $(T \times S, \mathcal{T} \otimes \mathcal{B}(S))$ whose projection on T equals μ, i.e.,

$$\lambda(A \times S) = \mu(A) \quad \text{for all } A \in \mathcal{T}.$$

An element of $\mathcal{Y}(T, \mu; S)$ is called a *Young measure.*

An important class of Young measures is the Young measure associated with a measurable function[3]. Let $f : T \longrightarrow S$ be measurable. Then the Young measure λ associated with the function f is defined by the image measure of μ by the map $t \longmapsto (t, f(t))$, i.e., λ is defined by the relation:

$$\lambda(A \times B) = \mu(A \cap f^{-1}(B)) \quad ; \quad A \in \mathcal{T}, \ B \in \mathcal{B}(S).$$

A measurable family of probabilities is called a *transition probability.* More precisely, $\nu : T \times \mathcal{B}(S) \longrightarrow \mathbb{R}$ is said to be a transition probability if the following two conditions are satisfied:

(i). for each $t \in T$, $\nu(t; \cdot)$ is a probability measure on S, and
(ii). for each $A \in \mathcal{B}(S)$, the function $t \longmapsto \nu(t; A)$ is measurable.

Let $\nu : T \times \mathcal{B}(S) \longrightarrow \mathbb{R}$ be a transition probability. We define the positive bounded measure λ on $T \times S$ by:

$$\lambda(A) = \int_T \left[\int_S \chi_A(t, s) \nu(t; ds) \right] \mu(dt) \quad ; \quad A \in \mathcal{T} \otimes \mathcal{B}(S). \tag{2.1}$$

Then λ is easily seen to be a Young measure.

The striking fact is that the inverse relation also holds. That is, for each Young measure, we can construct a transition probability which satisfies (2.1). In fact, the following proposition holds true.

---

[3] The following explanation of the Young measure associated with a function quoted from Valadier[21], p.353 may be helpful to understand this concept:

> The Young measure ν associated to u represents the amount of chalk (or ink) laid down when drawing the graph of u, but proportionally to the abscissa: $\nu(A \times \mathbb{R}^d) = \mu(A)$. Maybe a better analogy is the following: imagine a black television screen and that white light appears along the graph of u with intensity ν obeying to $\nu(A \times \mathbb{R}^d) = \mu(A)$.

Proposition 1 (Valadier[19]). *Let $(T, \mathcal{T}, \mu)$ be a complete finite positive measure space and S a Hausdorff topological space. Let λ be a positive bounded measure on $(T \times S, \mathcal{T} \otimes \mathcal{B}(S))$ whose projection on T equals μ and whose projection on S is Radon*[4]*. Then there exists a measurable family $\nu(t;\cdot)$ of Radon probabilities such that* (2.1) holds.

Remark that, since any bounded positive measure on a complete separable metric space is Radon, we can drop the condition "whose projection on S is Radon" in the above proposition. The transition probability ν derived in Proposition 1 is called the *disintegration of measure* λ. And Proposition 1 is known as the *disintegration theorem.*

Notice that, for each $\lambda \in \mathcal{Y}(T, \mu; S)$, there corresponds with many transition probabilities satisfying (2.1). However, there is exactly one-to-one correspondence between the space of Young measures and the equivalence class of μ-almost everywhere equal transition probabilities. This allows an identification of the two concepts. We identify, henceforth, the Young measure λ and its disintegration ν. Remark also that, for each λ-integrable function $\psi : T \times S \longrightarrow \mathbb{R}$,

$$\int_{T\times S} \psi d\lambda = \int_T \left[\int_S \psi(t,s)\nu(t;ds) \right] \mu(dt).$$

A function $\psi : T \times S \longrightarrow \mathbb{R}$ is said to be a *Carathéodory integrand* if the following three conditions hold:

(i). ψ is $\mathcal{T} \otimes \mathcal{B}(S)$-measurable,
(ii). for each $t \in T$, the map $s \longmapsto \psi(t,s)$ is continuous and bounded, and
(iii). $t \longmapsto \|\psi(t,\cdot)\|_\infty$ is μ-integrable[5],

where $\|\cdot\|_\infty$ stands for the sup-norm. The set of all Carathéodory integrands is denoted by $\mathcal{G}_C(T, \mu; S)$.

The set $\mathcal{Y}(T, \mu; S)$ is endowed with the weakest topology which makes the following linear functionals defined on $\mathcal{Y}(T, \mu; S)$ continuous:

$$\lambda \longmapsto \int_{T\times S} \psi d\lambda \quad (\psi \in \mathcal{G}_C(T, \mu; S)).$$

This topology on $\mathcal{Y}(T, \mu; S)$ is also called the *narrow topology.* Remark that the space $\mathcal{Y}(T, \mu; S)$ endowed with the narrow topology is Hausdorff

[4] A positive bounded Borel measure λ on S is said to be a *Radon measure* if, for any $A \in \mathcal{B}(S), \varepsilon > 0$, there exists a compact set $K \subset A$ such that $\lambda(A \setminus K) < \varepsilon$.

[5] In the usual usage of the term "Carathéodory integrand", only the conditions (i) and (ii) are postulated. In this paper, we impose an additional condition (iii).

and that, when S is compact, $\mathcal{Y}(T,\mu;S)$ is compact with respect to the narrow topology (see Valadier[20], Theorem 2 and Theorem 11).

A sequence of functions $\{f^n : S \longrightarrow \mathbb{R}\}$ is said to be *continuously convergent* to $f : S \longrightarrow \mathbb{R}$ at $x \in S$ if, for each sequence $\{x_n\}$ converging to x,

$$\lim_{n\to\infty} f^n(x_n) = f(x).$$

Proposition 2 (Maruyama[12]). *Let (S,d) be a complete separable metric space and λ^n, λ $(n \in \mathbb{N})$ be Borel probability measures on S such that*

$$\lambda^n \longrightarrow \lambda \ \textit{narrowly}.$$

Assume that $f^n \in \mathcal{C}^b(S)$ satisfy:

(i). *$\{f^n\}$ is uniformly bounded, and*

(ii). *$\{f^n\}$ is continuously convergent to f at all $x \in S$.*

Then,

$$\lim_{n\to\infty} \int_S f^n d\lambda^n = \int_S f d\lambda.$$

Proposition 3 (Valadier[20]). Let $(T_i, \mathcal{T}_i, \mu_i)$ $(i = 1,2)$ be a complete finite positive measure space, (S_i, d_i) a complete separable metric space, and λ_i an element of $\mathcal{Y}(T_i,\mu_i;S_i)$. We define $\lambda_1 \otimes \lambda_2 \in \mathcal{Y}(T_1 \times T_2, \mu_1 \otimes \mu_2; S_1 \times S_2)$ by

$$[\lambda_1 \otimes \lambda_2](t) = \lambda_1(t_1) \otimes \lambda_2(t_2) \ \ \textit{for each}\ t = (t_1,t_2) \in T_1 \times T_2.$$

Assume that the sequence $\{\lambda_i^n\}$ in $\mathcal{Y}(T_i,\mu_i;S_i)$ converges narrowly to λ_i. Then

$$\lambda_1^n \otimes \lambda_2^n \longrightarrow \lambda_1 \otimes \lambda_2 \ \ \textit{narrowly}.$$

Let $L(S,d)$ be the Banach space of real-valued Lipschitz functions defined on the compact metric space (S,d) endowed with the following norm:

$$\|f\|_L = \|f\|_\infty + \sup\left\{ \frac{|f(x) - f(y)|}{d(x,y)} : x \neq y \right\}.$$

Let us define, for each $\nu \in \mathcal{M}_+^1(S)$, a bounded linear functional $\Gamma_\nu : L(S,d) \longrightarrow \mathbb{R}$ by

$$\Gamma_\nu : f \longmapsto \int_S f d\nu.$$

Then we can imbed the space $\mathcal{M}_+^1(S)$ into the dual of $L(S,d)$ by the map $\nu \longmapsto \Gamma_\nu$. Furthermore, it is well-known that the narrow topology on $\mathcal{M}_+^1(S)$ coincides with the topology defined by the dual norm $\|\cdot\|_*$ on $L(S,d)^*$ and that $\mathcal{M}_+^1(S)$ is a closed convex subset of $L(S,d)^*$ when imbedded in this way (see Dudley[10], Theorem 11.3.3).

We denote by $\mathcal{C}(T,\mathcal{M}_+^1(S))$ the space of continuous functions $\lambda : T \longrightarrow \mathcal{M}_+^1(S)$ endowed with the uniform convergence norm:

$$\|\lambda\| = \sup_{t\in T} \|\lambda(t)\|_*.$$

Note that $\mathcal{C}(T,\mathcal{M}_+^1(S))$ can be interpreted as a class of transition probabilities. And so, by the disintegration theorem, it can also be regarded as a subset of $\mathcal{Y}(T,\mu;S)$.

3. Convergence of product Young measures

Let S_1, S_2 be complete separable metric spaces, and λ_1^n, λ_2^n $(n \in \mathbb{N} \cup \{\infty\})$ be Young measures on $T\times S_1, T\times S_2$ respectively. Let us define $\theta^n \in \mathcal{Y}(T,\mu;S_1\times S_2)$ by:

$$\theta^n(t) = \lambda_1^n(t)\otimes\lambda_2^n(t) \quad ; \; t\in T.$$

θ^n is called the *product Young measure* of λ_1^n and λ_2^n.

We consider, in this section, the convergence of product Young measures $\{\lambda_1^n\otimes\lambda_2^n\}$. It is well-known that the narrow convergence of $\{\lambda_1^n\otimes\lambda_2^n\}$ to $\lambda_1^\infty\otimes\lambda_2^\infty$ does not hold true even if we assume that each sequence $\{\lambda_i^n\}$ $(i=1,2)$ of Young measures converges narrowly to λ_i^∞ as the following example exemplified:

Example 2. Let $f_i^n : [0,1] \longrightarrow \mathbb{R}$ $(i=1,2)$ be the Rademacher functions defined by

$$f_i^n(x) = \begin{cases} 1 & \text{if } x \in [\frac{k}{2^n}, \frac{k+1}{2^n}] \text{ with } k \text{ even,} \\ -1 & \text{otherwise,} \end{cases}$$

and denote by δ_y the Dirac measure concentrated at y.

Since the Young measure[6] $\delta_{f_i^n(x)}$ associated with the Rademacher function $f_i^n(x)$ converges narrowly to $\frac{1}{2}(\delta_1+\delta_{-1})$ as $n \longrightarrow \infty$, the product of the narrow limits of $\{\delta_{f_i^n}(x)\}$ has the form:

[6] To be precise, the Young measure associated with the function $f(x)$ must be denoted by $\delta_x\otimes\delta_{f(x)}$ and the notation $\delta_{f(x)}$ represents the disintegration of the Young measure $\delta_x\otimes\delta_{f(x)}$. Since, by the disintegration theorem, we can identify the two concept, we use the notation $\delta_{f(x)}$ to represent also the Young measure associated with the function $f(x)$.

$$\frac{1}{4}(\delta_{(1,1)} + \delta_{(1,-1)} + \delta_{(-1,1)} + \delta_{(-1,-1)}).$$

On the other hand, the product Young measures $\{\delta_{f_1^n(x)} \otimes \delta_{f_2^n(x)}\}$ converges narrowly to the measure

$$\frac{1}{2}(\delta_{(1,1)} + \delta_{(-1,-1)}).$$

Thus,

$$\lim_{n\to\infty} \delta_{f_1^n(x)} \otimes \lim_{n\to\infty} \delta_{f_2^n(x)} \neq \lim_{n\to\infty} \delta_{f_1^n(x)} \otimes \delta_{f_2^n(x)}.$$

(This example is due to Valadier[21].)

And so, we assume that the sequence $\{\lambda_1^n(t)\}$ converges narrowly to $\lambda_1^\infty(t)$ in $\mathcal{M}_+^1(S_1)$ for all $t \in T$ in addition to the narrow convergence of $\{\lambda_2^n\}$ to λ_2^∞ in $\mathcal{Y}(T, \mu; S_2)$. Under these assumptions, the convergence of the product Young measures $\{\lambda_1^n \otimes \lambda_2^n\}$ to $\lambda_1^\infty \otimes \lambda_2^\infty$ holds true:

Theorem 1. *Let $(T, \mathcal{T}, \mu)$ be a complete finite positive measure space, S_1, S_2 be complete separable metric spaces and $\lambda_1^n \in \mathcal{Y}(T, \mu; S_1), \lambda_2^n \in \mathcal{Y}(T, \mu; S_2)$ $(n \in \mathbb{N} \cup \{\infty\})$. Let $\theta^n \in \mathcal{Y}(T, \mu; S_1 \times S_2)$ be the product Young measure of λ_1^n and λ_2^n defined, for each $t \in T$, by:*

$$\theta^n(t) = \lambda_1^n(t) \otimes \lambda_2^n(t).$$

Assume that:

(i). *$\lambda_1^n(t) \longrightarrow \lambda_1^\infty(t)$ narrowly in $\mathcal{M}_+^1(S_1)$ for all $t \in T$, and*
(ii). *$\lambda_2^n \longrightarrow \lambda_2^\infty$ narrowly in $\mathcal{Y}(T, \mu; S_2)$.*

Then

$$\theta^n \longrightarrow \theta^\infty \quad \textit{narrowly in } \mathcal{Y}(T, \mu; S_1 \times S_2).$$

Proof of Theorem 1. The proof is essentially the same as Valadier[20], Theorem 14. (See Corollary 2 of the present paper.) We have to show that

$$\int_{T\times S_1\times S_2} \psi d\theta^n \longrightarrow \int_{T\times S_1\times S_2} \psi d\theta^\infty$$

for any $\psi \in \mathcal{G}_C(T, \mu; S_1 \times S_2)$.

Step 1. Let $\hat{\mathbb{N}} = \mathbb{N} \cup \{\infty\}$ be the Alexandrov one point compactification of the space of natural numbers $\mathbb{N}$. We define $\overline{\psi} : T \times (\hat{\mathbb{N}} \times S_2) \longrightarrow \mathbb{R}$ by

$$\overline{\psi}(t, k, y) = \int_{S_1} \psi(t, x, y)\lambda_1^k(t)(dx).$$

Then it is easily verified that $\overline{\psi}$ is a Carathéodory integrand as follows. To start with, we can check the measurability of the map $t \longmapsto \overline{\psi}(t, k, y)$ according to Valadier[20], Lemma A2. Furthermore, since

$$\|\overline{\psi}(t,\cdot,\cdot)\|_\infty \leqq \|\psi(t,\cdot,\cdot)\|_\infty$$

and $\hat{\mathbb{N}} \times S_2$ is separable, the μ-integrability of $t \longmapsto \|\overline{\psi}(t,\cdot,\cdot)\|_\infty$ is a direct consequence of the μ-integrability of $t \longmapsto \|\psi(t,\cdot,\cdot)\|_\infty$ as soon as the continuity of the map $(k, y) \longmapsto \overline{\psi}(t, k, y)$ can be shown. And this is what we have to do next.

Step 2. Let $\{(k_p, y_p)\}$ be a sequence in $\hat{\mathbb{N}} \times S_2$ which converges to $(k_\infty, y_\infty) \in \hat{\mathbb{N}} \times S_2$. If $k_\infty \in \mathbb{N}$, then $k_p = k_\infty$ for sufficiently large p. Thus, by Lebegsue's bounded convergence theorem, we obtain

$$\overline{\psi}(t, k_p, y_p) \longrightarrow \overline{\psi}(t, k_\infty, y_\infty).$$

Step 3. This step is devoted to the case $k_\infty = \infty$. Let $\psi^p(t, x)$ be defined by

$$\psi^p(t, x) = \psi(t, x, y_p).$$

Then, for any $x_\infty \in S_1$ and any sequence $\{x_p\}$ in S_1 which converges to x_∞,

$$\psi^p(t, x_p) = \psi(t, x_p, y_p) \longrightarrow \psi(t, x_\infty, y_\infty) = \psi^\infty(t, x_\infty),$$

that is, for each fixed $t \in T$, the sequence of functions $\{x \longmapsto \psi^p(t, x)\}$ is continuously convergent to ψ^∞ at all $x_\infty \in S_1$. And we deduce, by Proposition 2 and Assumption (i), that:

$$\overline{\psi}(t, k_p, y_p) \longrightarrow \overline{\psi}(t, \infty, y_\infty).$$

Step 4. Let $T_1 = \{t_1\}$ be a set consisting of one point t_1 and define the probability measure μ_1 on T_1 by $\mu_1(\{t_1\}) = 1$. Then, by Proposition 3 and Assumption (ii):

$$\begin{aligned}
&\int_{T\times S_1\times S_2} \psi d\theta^n \\
&= \int_T \left[\int_{S_1\times S_2} \psi(t,x,y)(\lambda_1^n(t)\otimes\lambda_2^n(t))(d(x,y))\right]\mu(dt) \\
&= \int_T \left[\int_{\hat{\mathbb{N}}\times S_2} \overline{\psi}(t,k,y)(\delta_n\otimes\lambda_2^n(t))(d(k,y))\right]\mu(dt) \\
&= \int_{T_1\times T} \left[\int_{\hat{\mathbb{N}}\times S_2} \overline{\psi}(t,k,y)(\delta_n\otimes\lambda_2^n(t))(d(k,y))\right](\mu_1\otimes\mu)(d(t_1,t)) \\
&\longrightarrow \int_{T_1\times T} \left[\int_{\hat{\mathbb{N}}\times S_2} \overline{\psi}(t,k,y)(\delta_\infty\otimes\lambda_2^\infty(t))(d(k,y))\right](\mu_1\otimes\mu)(d(t_1,t))
\end{aligned}$$

$$= \int_T \left[\int_{S_2} \overline{\psi}(t, \infty, y) \lambda_2^\infty(t)(dy) \right] \mu(dt)$$
$$= \int_{T \times S_1 \times S_2} \psi d\theta^\infty.$$

Remark that Assumption (i) of the pointwise narrow convergence implies the narrow convergence of Young measures thanks to Lebesgue's dominated convergence theorem. And thus, the following corollary is the direct consequence of Theorem 1.

Corollary 1. *Let $(T, \mathcal{T}, \mu)$ be a complete finite positive measure space, S_1, S_2 be complete separable metric spaces and $\lambda_1^n \in \mathcal{Y}(T, \mu; S_1), \lambda_2^n \in \mathcal{Y}(T, \mu; S_2)$ $(n \in \mathbb{N} \cup \{\infty\})$. Let $\theta^n \in \mathcal{Y}(T, \mu; S_1 \times S_2)$ be defined as in Theorem 1. Assume that, for each $i = 1, 2$,*

$$\lambda_i^n(t) \longrightarrow \lambda_i^\infty(t) \ \textit{narrowly in} \ \mathcal{M}_+^1(S_i) \ \textit{for all} \ t \in T.$$

Then

$$\theta^n \longrightarrow \theta^\infty \ \textit{narrowly in} \ \mathcal{Y}(T, \mu; S_1 \times S_2).$$

Corollary 2 (Valadier[20], Theorem 14). *Let $(T, \mathcal{T}, \mu)$ be a complete finite positive measure space, S_1, S_2 be complete separable metric spaces, $f^n : T \longrightarrow S_1$ be measurable and λ^n be in $\mathcal{Y}(T, \mu; S_2)$. We define $\theta^n \in \mathcal{Y}(T, \mu; S_1 \times S_2)$ by*

$$\theta^n(t) = \delta_{f^n(t)} \otimes \lambda^n(t).$$

Assume that

(i). $f^n \longrightarrow f^\infty$ *in measure, and*
(ii). $\lambda^n \longrightarrow \lambda^\infty$ *narrowly.*

Then

$$\theta^n \longrightarrow \theta^\infty \quad \textit{narrowly}.$$

Proof. Assume, on the contrary, that $\{\theta^n\}$ does not converge narrowly to θ^∞. Then there exist a subsequence $\{\theta^{n_k}\}$, $\varepsilon > 0$, and a Caratéodory integrand $\psi(t, x, y)$ such that

$$\left| \int_{T \times S_1 \times S_2} \psi d\theta^{n_k} - \int_{T \times S_1 \times S_2} \psi d\theta^\infty \right| > \varepsilon \ \text{ for all } n_k.$$

Since, by Assumption (i), $\{f^{n_k}\}$ converges in measure to f^∞, by taking further subsequence if necessary, we may assume that $\{f^{n_k}\}$ converges almost everywhere to f^∞. By identifying μ-almost everywhere equal functions, we can apply Theorem 1. And thus $\{\theta^{n_k}\}$ converges narrowly to θ^∞. This is a contradiction.

4. Existence of equilibria of an equicontinuous game: the case of compact metric space of types

In this section, we consider the existence of equilibria of equicontinuous games in the case that each player's type set is a compact metric space.

For simplicity, assume that there are two players $i = 1, 2$. Let (T_i, ρ_i) be a compact metric space of types and (S_i, d_i) be a compact metric space of actions for each $i = 1, 2$. We set $T = T_1 \times T_2$. Assume that T is equipped with a regular probability measure μ defined on the σ-algebra $\mathcal{T}$ and let μ_i $(i = 1, 2)$ be the marginal probability of μ on T_i and $\mathcal{T}_i$ be the σ-algebra on T_i on which μ_i is defined.

Suppose that, for each $i = 1, 2$, a function $u_i : T \times S_i \times S_j \longrightarrow \mathbb{R}$ is given and define each player's payoff function $U_i : \mathcal{Y}(T_i, \mu_i; S_i) \times \mathcal{Y}(T_j, \mu_j; S_j) \longrightarrow \mathbb{R}$ as follows:

$$U_i(\lambda_i, \lambda_j) = \int_T \left[\int_{S_i} \int_{S_j} u_i(t; s_i, s_j) \lambda_i(t_i; ds_i) \lambda_j(t_j; ds_j) \right] \mu(dt), \tag{4.1}$$

where $\{j\} = \{1, 2\} \setminus \{i\}$ and $t = (t_1, t_2) \in T_1 \times T_2$. Furthermore, let $\mathcal{S}_i \subset \mathcal{C}(T_i, \mathcal{M}_+^1(S_i))$ satisfy the following conditions:

(i). $\mathcal{S}_i$ is convex, and closed with respect to the uniform convergence topology.

(ii). $\mathcal{S}_i$ is equicontinuous, i.e., for any $x \in T_i$ and $\varepsilon > 0$, there exists $\delta(x) > 0$ such that

$$\rho_i(x, y) < \delta(x), \ y \in T_i \Longrightarrow \|\lambda(x; \cdot) - \lambda(y; \cdot)\|_* \leqq \varepsilon \quad \text{for all } \lambda \in \mathcal{S}_i.$$

We call $\mathcal{S}_i$ an *equicontinuous strategy set* for player i and the Bayesian game in which each player's strategy set is equicontinuous an *equicontinuous game.*

Example 3. Let

$$\mathcal{S}_i = \{\lambda \in \mathcal{Y}(T_i, \mu_i; S_i) : \|\lambda(x; \cdot) - \lambda(y; \cdot)\|_* \leqq K_i \cdot \rho_i(x, y), \ x, y \in T_i\},$$

where K_i $(i = 1, 2)$ is a given constant. Then $\mathcal{S}_i$ is an equicontinuous strategy set for player i.

Then we have the following theorem:

Theorem 2. *Let T_i, S_i be compact metric spaces, and $\mathcal{S}_i$ be an equicontinuous strategy set for $i = 1, 2$. Assume that $u_i : T \times S_i \times S_j \longrightarrow \mathbb{R}$ $(i = 1, 2)$ satisfy the following conditions:*

(i). *u_i is $\mathcal{T} \otimes \mathcal{B}(S_i) \otimes \mathcal{B}(S_j)$-measurable,*

(ii). *for each* $t \in T$, *the map* $(s_i, s_j) \longmapsto u_i(t; s_i, s_j)$ *is continuous, and*
(iii). *there exists a* μ*-integrable real-valued function* $\phi(t)$ *such that*

$$|u_i(t; s_i, s_j)| \leqq \phi(t) \quad \textit{for all } (t, s_i, s_j) \in T \times S_i \times S_j.$$

Then there exists a pair of strategies $(\lambda_1^*, \lambda_2^*) \in \mathcal{S}_1 \times \mathcal{S}_2$ *such that*

$$U_i(\lambda_i, \lambda_j^*) \leqq U_i(\lambda_i^*, \lambda_j^*) \quad \textit{for all } \lambda_i \in \mathcal{S}_i.$$

We shall prove Theorem 2 by making use of the following proposition:

Proposition 4 (Browder[5]). *Let* C_1, C_2 *be nonempty compact convex sets each in a real topological space. Assume that the function* $\pi_i : C_i \times C_j \longrightarrow \mathbb{R}$ *satisfy the following conditions for* $i = 1, 2$:

(i). π_i *is continuous, and*
(ii). *for each* $c_j \in C_j$, *the map* $c_i \longmapsto \pi_i(c_i, c_j)$ *is quasi-concave.*

Then there exists a pair of points $(c_1^*, c_2^*) \in C_1 \times C_2$ *such that*

$$\pi_i(c_i, c_j^*) \leqq \pi_i(c_i^*, c_j^*) \quad \textit{for all } c_i \in C_i.$$

Remark 1. Since the payoff function (4.1) is linear in each variable, it is certainly quasi-concave. Furthermore, since S_i is compact, $\mathcal{Y}(T_i, \mu_i; S_i)$ is compact with respect to the narrow topology. Thus, in applying the above proposition to the existence of equilibria of a Bayesian game, only the continuity condition (i) of the payoff function is the problem. If we assume the regularity condition: μ is absolutely continuous with respect to $\mu_1 \otimes \mu_2$, the payoff function (4.1) can be rewritten as:

$$U_i(\lambda_i, \lambda_j) = \int_T \left[\int_{S_i} \int_{S_j} g(t) u_i(t; s_i, s_j) \lambda_i(t_i; ds_i) \lambda_j(t_j; ds_j) \right] \mu_1 \otimes \mu_2(dt),$$

where $g(t)$ is the Radon-Nikodým derivative of μ with respect to $\mu_1 \otimes \mu_2$. And thus, by applying Proposition 3, we can easily check the continuity of the payoff function (4.1).

Proof of Theorem 2. By embedding $\mathcal{M}_+^1(S_i)$ into the space $L(S_i, d_i)^*$, $\mathcal{S}_i$ can be regarded as a set of functions taking values in the Banach space $L(S_i; d_i)^*$. Since $\mathcal{S}_i$ is an equicontinuous family of functions, by applying Ascoli's Theorem (Dieudonné[8], (7.5.7)), we deduce that $\mathcal{S}_i$ is compact with respect to the uniform convergence topology. Furthermore, by Corollary 1, the payoff function $U_i : \mathcal{S}_i \times \mathcal{S}_j \longrightarrow \mathbb{R}$ is continuous with respect to this topology. Hence, by applying Proposition 4, we deduce Theorem 2.

5. Some lemmas

In this section, we shall provide some lemmas used, in the next section, to establish the ε-approximate equilibria of an equicontinuous game in which each player's type set is a separable metric space. And so, in this and the next sections, we assume that (T_i, ρ_i) be a separable metric space of types. The other settings and assumptions are the same as in section 4.

Let $\eta_i : T_i \times \mathcal{T}_i \longrightarrow [0,1]$ be such that

(i). for each $A \in \mathcal{T}_i$, $t \longmapsto \eta_i(t; A)$ is measurable,
(ii). for each $t \in T_i$, $\eta_i(t; \cdot)$ is a probability measure on T_i, and
(iii). for each $t \in T_i$, $\eta_i(t; \cdot)$ is absolutely continuous with respect to μ_i.

We call η_i which satisfies the above three conditions *a measure of approximation of player i*.

Remark that, for the Young measure $\lambda \in \mathcal{Y}(T_i, \mu_i; S_i)$, the following two conditions are equivalent (see Valadier[20], Lemma A2):

(i). for each $A \in \mathcal{B}(S_i)$, the function $t \longmapsto \lambda(t, A)$ is measurable, and
(ii). for each $\varphi \in \mathcal{C}(S_i)$, the function $t \longmapsto \int_{S_i} \varphi(s)\lambda(t, ds)$ is measurable.

Taking this fact into accout, each $\lambda \in \mathcal{Y}(T_i, \mu_i; S_i)$ can be regarded as a measurable function defined on T_i into the Banach space $L(S_i, d_i)^*$. And thus the following Bochner integral is well-defined:

$$\lambda^{\eta_i}(t; \cdot) = \int_{T_i} \lambda(y; \cdot)\eta_i(t; dy).$$

And we have the following lemmas[7].

Lemma 1. *The function $\lambda^{\eta_i} : T_i \longrightarrow L(S_i, d_i)^*$ constructed above defines a Young measure and satisfies the following relation:*

$$\lambda^{\eta_i}(t; A) = \int_{T_i} \lambda(y; A)\eta_i(t; dy) \quad ; \quad A \in \mathcal{B}(S_i). \tag{5.1}$$

Proof. The relation (5.1) implies that $\lambda^{\eta_i}(t; \cdot)$ is a probability measure on S_i. Furthermore, the measurability of $t \longmapsto \lambda^{\eta_i}(t; A)$; $A \in \mathcal{B}(S_i)$ follows from (5.1) and Neveu[18], Proposition 3.2.1. And so we need only to check the relation (5.1). To start with, let F be any closed subset of S_i and define $\varphi_n : S_i \longrightarrow \mathbb{R}$ by

[7] The proofs of these lemmas are improved considerably by the comments from an anonymous referee.

$$\varphi_n(s) = [1 - nd_i(s, F)]^+,$$

where $[a]^+ = \max\{0, a\}$ and $d_i(s, F) = \inf_{x \in F} d_i(s, x)$. Then φ_n is easily seen to be a Lipschitz function. Hence,

$$\int_{S_i} \varphi_n(s) \lambda^{\eta_i}(t; ds) = \int_{T_i} \int_{S_i} \varphi_n(s) \lambda(y; ds) \eta_i(t; dy).$$

Since $\varphi_n(s) \downarrow \chi_F(s)$ as $n \longrightarrow \infty$, it follows that

$$\lambda^{\eta_i}(t; F) = \int_{T_i} \lambda(y; F) \eta_i(t; dy).$$

And thus, we conclude that (5.1) holds true.

Lemma 2. *Let $f : S_i \longrightarrow \mathbb{R}$ be a bounded measurable function. Then:*

$$\int_{S_i} f(s) \lambda^{\eta_i}(t; ds) = \int_{T_i} \left[\int_{S_i} f(s) \lambda(y; ds) \right] \eta_i(t; dy).$$

Proof. This lemma follows immediately from (5.1).

Example 4. Let us assume that the Radon-Nikodým derivative $g_i(t, y)$ of η_i with respect to μ_i be integrably bounded, and continuous in t. Then the function λ^{η_i} defined on T_i into $\mathcal{M}_+^1(S_i)$ is continuous. In fact, for any continuous function $f : S_i \longrightarrow \mathbb{R}$, $t \in T$, and any sequence $\{t_n\}$ converging to t, the following estimations can be obtained:

$$\begin{aligned} &\int_{S_i} f(s)[\lambda^{\eta_i}(t_n; ds) - \lambda^{\eta_i}(t; ds)] \\ &= \int_{T_i} \left[\int_{S_i} f(s) \lambda(y; ds) \right] [\eta_i(t_n; dy) - \eta_i(t; dy)] \\ &= \int_{T_i} (g_i(t_n, y) - g_i(t, y)) \left[\int_{S_i} f(s) \lambda(y; ds) \right] \mu_i(dy), \end{aligned}$$

where the first equality is the consequence of Lemma 2. Since the function $y \longmapsto \int_{S_i} f(s) \lambda(y; ds)$ is bounded, and $g_i(t_n, \cdot) \longrightarrow g_i(t, \cdot)$ in $\mathcal{L}_1(T_i, \mathcal{T}_i, \mu_i)$ as $n \longrightarrow \infty$ by virtue of the Lebesgue dominated convergence theorem, the right-hand side of the above equality tends to 0 as $n \longrightarrow \infty$.

Lemma 3. *Let $\{\lambda_n\}$ be a sequence in $\mathcal{Y}(T_i, \mu_i; S_i)$ which converges narrowly to $\lambda \in \mathcal{Y}(T_i, \mu_i; S_i)$. Then the following convergence holds: for each $t \in T_i$,*

$$\lambda_n^{\eta_i}(t) \longrightarrow \lambda^{\eta_i}(t) \text{ narrowly in } \mathcal{M}_+^1(S_i).$$

Proof. Let f be a continuous real-valued function defined on S_i. Then, by Lemma 2:

$$\begin{aligned}\int_{S_i} f(s)\lambda_n^{\eta_i}(t;ds) &= \int_{T_i}\left[\int_{S_i} f(s)\lambda_n(y;ds)\right]\eta_i(t;dy) \\ &\longrightarrow \\ &\int_{T_i}\left[\int_{S_i} f(s)\lambda(y;ds)\right]\eta_i(t;dy) \\ &= \int_{S_i} f(s)\lambda^{\eta_i}(t;ds).\end{aligned}$$

6. Existence of ε-approximate equilibria of an equicontinuous game : the case of separable metric space of types

In this section, we consider an equicontinuous game examined in section 4 with generalization that the type set of each player is a separable metric space and deduce the existence of ε-approximate equilibria. The main theorem shown in this section is as follows:

Theorem 6.1. *Let T_i be a separable metric space, S_i a compact metric space, and $\mathcal{S}_i$ an equicontinuous strategy set for $i = 1, 2$. Assume that $u_i : T \times S_i \times S_j \longrightarrow \mathbb{R}$ $(i = 1, 2)$ satisfy the following conditions:*

(i) *$u_i(t, s_i, s_j)$ is $\mathcal{T} \otimes \mathcal{B}(S_i) \otimes \mathcal{B}(S_j)$-measurable,*
(ii) *for each $t \in T$, the map $(s_i, s_j) \longmapsto u_i(t; s_i, s_j)$ is continuous,*
(iii) *there exists M_i such that*

$$|u_i(t; s_i, x) - u_i(t; s_i, y)| \leqq M_i \cdot d_j(x, y)$$

for all $(t, s_i) \in T \times S_i$, and
(iv) *there exists μ-integrable real-valued function $\phi(t)$ such that*

$$|u_i(t; s_i, s_j)| \leqq \phi(t) \quad \text{for all } (t, s_i, s_j) \in T \times S_i \times S_j.$$

Then, for each $\varepsilon > 0$, there exists a pair of strategies $(\lambda_1^*, \lambda_2^*) \in \mathcal{S}_1 \times \mathcal{S}_2$ such that

$$U_i(\lambda, \lambda_j^*) \leqq U_i(\lambda_i^*, \lambda_j^*) + \varepsilon \quad \text{for all } \lambda \in \mathcal{S}_i.$$

In order to prove the theorem, we introduce, a function $P_i : \mathcal{Y}(T_i, \mu_i; S_i) \times \mathcal{Y}(T_j, \mu_j; S_j) \longrightarrow \mathbb{R}$ defined by:

$$P_i(\lambda_i, \lambda_j; \eta_j) = \int_T \left[\int_{S_i}\int_{S_j} u_i(t; s_i, s_j)\lambda_i(t_i; ds_i)\lambda_j^{\eta_j}(t_j; ds_j)\right]\mu(dt),$$

where η_j is a measure of approximation of player j. Furthermore, we denote by $\overline{\mathcal{S}_i}$ the closure of $\mathcal{S}_i$ with respect to the narrow topology in $\mathcal{Y}(T_i, \mu_i; S_i)$.

Lemma 4. *Let T_i be a separable metric space, S_i a compact metric space, $\mathcal{S}_i$ an equicontinuous strategy set, and η_i a measure of approximation of player i for $i = 1, 2$. Suppose that $u_i : T \times S_i \times S_j \longrightarrow \mathbb{R}$ satisfy the conditions of Theorem 3. Then there exists a pair of Young measures $(\lambda_1^*, \lambda_2^*) \in \overline{\mathcal{S}_1} \times \overline{\mathcal{S}_2}$ such that*

$$P_i(\lambda_i, \lambda_j^*; \eta_j) \leqq P_i(\lambda_i^*, \lambda_j^*; \eta_j) \quad \text{for all } \lambda_i \in \overline{\mathcal{S}_i}$$

Proof. Let us set $C_i = \overline{\mathcal{S}_i}, \pi_i = P_i$ in Proposition 4. Since $\mathcal{Y}(T_i, \mu_i; S_i)$ is compact with respect to the narrow topology, $\overline{\mathcal{S}_i}$ is compact as the closed subset of a compact set. Furthermore, the continuity of P_i is implied by Theorem 1 and Lemma 3. Also, since the map $\lambda_i \longmapsto P_i(\lambda_i, \lambda_j; \eta_j)$ is linear for each λ_j, it is certainly quasi-concave. And thus, by applying Proposition 4, we conclude that the assertion holds true.

Lemma 5. *Let T_i be a separable metric space, S_i a compact metric space, $\mathcal{S}_i$ an equicontinuous strategy set, and η_i a measure of approximation of player i for $i = 1, 2$. Suppose that $u_i : T \times S_i \times S_j \longrightarrow \mathbb{R}$ satisfy the conditions of Theorem 3. Then, for each $\varepsilon > 0$, there exists a pair of strategies $(\lambda_1^*, \lambda_2^*) \in \mathcal{S}_1 \times \mathcal{S}_2$ such that*

$$P_i(\lambda_i, \lambda_j^*; \eta_j) \leqq P_i(\lambda_i^*, \lambda_j^*; \eta_j) + \varepsilon \quad \text{for all} \quad \lambda_i \in \mathcal{S}_i.$$

Proof. Let $(\lambda_1^*, \lambda_2^*) \in \overline{\mathcal{S}_1} \times \overline{\mathcal{S}_2}$ be a pair of Young measures obtained in Lemma 4 and $(\lambda_1^n, \lambda_2^n) \in \mathcal{S}_1 \times \mathcal{S}_2$ be a sequence converging to $(\lambda_1^*, \lambda_2^*)$ with respect to the narrow topology. Since the map

$$(\lambda_i, \lambda_j) \longmapsto P_i(\lambda_i, \lambda_j; \eta_j) - \max_{\lambda \in \overline{\mathcal{S}_i}} P_i(\lambda, \lambda_j; \eta_j)$$

is continuous with respect to the narrow topology by Theorem 1, Lemma 3 and Berge's maximum theorem, we conclude that, for sufficiently large n,

$$P_i(\lambda_i^n, \lambda_j^n; \eta_j) - \max_{\lambda \in \overline{\mathcal{S}_i}} P_i(\lambda, \lambda_j^n; \eta_j) \geqq -\varepsilon.$$

Thus, for such n, the pair $(\lambda_1^n, \lambda_2^n)$ satisfies the assertion.

Proof of Theorem 3. Let $r_j > 0$ be such that

$$r_j < \frac{\varepsilon}{3\left(\int_T \phi d\mu + M_i\right)}$$

and define $\delta_j(x)$ by

$$\delta_j(x) = \inf\{\rho_j(x,y) : \sup_{\lambda \in \mathcal{S}_j} \|\lambda(x;\cdot) - \lambda(y;\cdot)\|_* > r_j\}.$$

Then, it is easy to verify that the function $\delta_j(x)$ is measurable as follows. First, we define $\Gamma : T_j \longrightarrow\!\!\!\!\rightarrow T_j$ by

$$\Gamma(x) = \left\{ y \in T_j : \sup_{\lambda \in \mathcal{S}_j} \|\lambda(x;\cdot) - \lambda(y;\cdot)\|_* > r_j \right\}.$$

Then, since

$$\mathrm{Graph}(\Gamma) = \bigcup_{\lambda \in \mathcal{S}_j} \{(x,y) : \|\lambda(x;\cdot) - \lambda(y;\cdot)\|_* > r_j\},$$

Graph(Γ) is open and thus, Γ is measurable. Hence the measurability of δ_j is implied by Aubin-Frankowska [2], Theorem 8.2.11 and the fact that δ_j is rewritten as:

$$\delta_j(x) = \inf_{y \in \overline{\Gamma}(x)} \rho_j(x,y),$$

where $\overline{\Gamma}$ stands for the closure of its values.

Notice that, since $\mathcal{S}_j$ is an equicontinuous family, $\delta_j(x) > 0$ for all $x \in T_j$.

Let us define $\eta_j(t;A)$, $(t,A) \in T_j \times \mathcal{T}_j$ by

$$\eta_j(t;A) = \begin{cases} \dfrac{\mu_j(B_{\delta_j(t)}(t) \cap A)}{\mu_j(B_{\delta_j(t)}(t))} & \text{if } t \in \text{ support } \mu_j \\ \mu_j(A) & \text{if } t \notin \text{ support } \mu_j, \end{cases}$$

where $B_{\delta_j(t)}(t)$ stands for the open ball with center at t and radius $\delta_j(t)$. Then, since, for each fixed $A \in \mathcal{T}_j$, the graph of $t \longmapsto\!\!\!\!\rightarrow B_{\delta_j(t)}(t) \cap A$ is measurable, the map $t \longmapsto \mu_j(B_{\delta_j(t)}(t) \cap A)$ is measurable by virtue of Fubini's Theorem. And thus η_j is a measure of approximation of player j. Applying Lemma 5, we obtain a pair of strategies $(\lambda_1^*, \lambda_2^*) \in \mathcal{S}_1 \times \mathcal{S}_2$ such that

$$U_i(\lambda_i, \lambda_j^{*\eta_j}) \leqq U_i(\lambda_i^*, \lambda_j^{*\eta_j}) + \frac{\varepsilon}{3} \quad \text{for all } \lambda_i \in \mathcal{S}_i.$$

Furthermore, for each $\lambda_i \in \mathcal{S}_i$, the following estimations can be obtained:

$$
\begin{aligned}
&|U_i(\lambda_i, \lambda_j^{*\eta_j}) - U_i(\lambda_i, \lambda_j^*)| \\
&\leqq \int_T \int_{S_i} \left| \int_{S_j} u_i(t; s_i, s_j)[\lambda_j^{*\eta_j}(t_j; ds_j) - \lambda_j^*(t_j; ds_j)] \right| \lambda_i(t_i; ds_i)\mu(dt) \\
&\leqq \int_T \int_{S_i} \left(\sup_{s_j} |u_i(t; s_i, s_j)| + M_i \right) \|\lambda_j^{*\eta_j}(t_j; \cdot) - \lambda_j^*(t_j; \cdot)\|_* \lambda_i(t_i; ds_i)\mu(dt) \\
&\leqq \int_T (\phi(t) + M_i) \left\| \int_{T_j} [\lambda_j^*(y; \cdot) - \lambda_j^*(t_j; \cdot)]\eta_j(t_j; dy) \right\|_* \mu(dt) \\
&\leqq \int_T (\phi(t) + M_i) \int_{T_j} \|\lambda_j^*(y; \cdot) - \lambda_j^*(t_j; \cdot)\|_* \eta_j(t_j; dy)\mu(dt) \\
&\leqq \int_T (\phi(t) + M_i) \int_{T_j} r_j \cdot \eta_j(t_j; dy)\mu(dt) \\
&\leqq \frac{\varepsilon}{3}.
\end{aligned}
$$

Thus, we conclude that

$$
U_i(\lambda_i, \lambda_j^*) \leqq U_i(\lambda_i^*, \lambda_j^*) + \varepsilon \quad \text{for all } \lambda_i \in \mathcal{S}_i.
$$

References

1. Ambruster, W., Böge, W.: Bayesian game theory. In: Moeschlin, O. et al. (eds) Game Theory and Related Topics. North-Holland, Amsterdam 1979
2. Aubin, J.-P., Frankowska, H.: Set-valued Analysis. Birkhäuser, Boston 1990
3. Balder, E.J.: Generalized equilibrium results for games with incomplete information. Math. Oper. Res. **13**, 265-276 (1988)
4. Balder, E.J.: A unifying approach to existence of Nash equilibria. International J. Game Theory. **24**, 79-94 (1995)
5. Browder, F.E.: The fixed point theory of multi-valued mappings in topological vector spaces. Math. Ann. **177**, 283-301 (1968)
6. Castaing, C., Valadier, M.: Weak convergence using Young measures. Functiones et Approximatio **26**, 7-17 (1998)
7. Diestel, J., Uhl, J.J.: Vector Measures. Amer. Math. Soc., Providence 1977
8. Dieudonné, J.: Foundations of Modern Analysis. Academic Press, N.Y. 1969
9. Dinculeanu, N.: Vector Measures. Pergamon Press, London 1967
10. Dudley, R.M.: Real Analysis and Probability. Wadsworth & Brooks/Cole Mathematics series, California 1989
11. Harsanyi, J.C.: Games with incomplete information played by "Bayesian" players. Management Science **14**, 159-182, 320-334, 486-502 (1967-68)
12. Maruyama, T.: A note on the disintegration of measures: a convergence theorem. Keio Economic Studies **16**, 57-64 (1979)

13. Khan, M.A.: On extensions of the Cournot-Nash theorem. In: Aliprantis, C.D. et al. (eds.): Advances in Equilibrium Theory, (Lecture Notes in Economics and Mathematical Systems Vol.244), pp. 79-106, Springer-Verlag, Berlin 1985
14. Mamer, J.W., Schilling, K.E.: A zero-sum game with incomplete information and compact action spaces. Math. Oper. Res. **11**, 627-631 (1986)
15. Mertens, J.-F., Zamir, S.: Formulation of Bayesian analysis for games with incomplete information. International J. Game Theory **14**, 1-29 (1985)
16. Milgrom, P.R., Weber, R.J.: Distributional strategies for games with incomplete information. Math. Oper. Res. **10**, 619-632 (1985)
17. Myerson, R.B.: Game Theory. Harvard University Press, Cambridge 1991
18. Neveu, J.: Mathematical Foundations of the Calculus of Probability. Holden-Day, San Francisco 1965
19. Valadier, M.: Désintégration d'une mesure sur un produit. C. R. Acad. Sci. Paris **A276**, 33-35 (1973)
20. Valadier, M.: Young measures. In: Cellina, A. (eds) Methods of Nonconvex Analysis (Lecture Notes in Mathematics Vol.1446). pp. 152-188, Springer-Verlag, Berlin 1989
21. Valadier, M.: A course on Young measures. In: Workshop di Teoria della Misura e Analisi Reale, Grado, September 19-October 2 1993. Rend. Matematica Trieste **26**(suppl.), 349-394 (1994)
22. Young, L.C.: Generalized curves and the existence of an attained absolute minimum in the calculus of variations. Comptes Rendus de la Société des Sc. et des lettres de Varsovie cl. III 30, 212-234 (1937)

Adv. Math. Econ. 4, 61–72 (2002)

Advances in
MATHEMATICAL
ECONOMICS

A Bayesian equilibrium existence theorem

Nicholas C. Yannelis*

Department of Economics, University of Illinois at Urbana-Champaign, Champaign, IL 61820, USA

Received: June 7, 2001
Revised: September 17, 2001

JEL classification: C72, C11

Mathematics Subject Classification (2000): 46 E 30, 28 A 20, 60 B 12

Abstract. We generalize the social equilibrium existence theorem of Debreu (1952) to a Bayesian setting. Our new Bayesian result includes as a special case not only the existence results of Debreu (1952) and Nash (1951) but also the recent Bayesian equilibrium existence result of Kim-Yannelis (1997).

Key words: Bayesian abstruct economy, Bayesian equilibrium, Bochner integration, Fatou's lemma in infinite dimensions, weak compactness

1. Introduction

The equilibrium existence result of Nash (1951) for a normal form game, was generalized by Debreu (1952) to a social system or abstract economy. The abstract economy was the main tool to prove the existence of a Walrasian equilibrium in concrete economy as Arrow-Debreu (1954) showed. There have been several generalization of the Nash equilibrium existence theorems to a Bayesian framework (see Kim-Yannelis (1997)) for an extensive list of references).

The propose of this paper is to present a new generalization of the Debreu (1952) existence theorem for a Bayesian abstract economy (or social system). This theorem generalizes the results of Debreu (1952), Nash (1951), Kim-Yannelis (1997) and Yannelis [1998], also, extends a recent related result of Aliprantis-Tourky-Yannelis (2001).

We would like to comment on the difficulties that one faces in order to generalize the theorems of Debreu [1952] and Nash [1951] to a Bayesian

* I wish to thank the Editor and the referees for useful comments

abstract economy. There are two main technical obstacles that one must overcome. First, the strategy set is a random set-valued correspondence with weakly compact, convex values, and we are working with the set of all private information measurable and Bochner integrable selections from that set. The latter set must be shown that it is weakly compact. This is done by using an extension of Diestel's theorem on weak compactness (see Theorem 3.1 in Yannelis [1991b]). Secondly, one must show that the set of all Bochner integrable selections from an upper semicontinuous, (u.s.c.) correspondence, is also u.s.c. This result is similar to the one that integration preserves u.s.c. and one way to prove this theorem is by means of the Fatou Lemma in infinite dimensions (e.g., Yannelis [1991b, Section 5]). In addition to the two above main difficulties there are also other secondary technical details entailing to the measurability properties of certain set-valued functions. Those details are not routine and the Aumann measurable selection theorem, and the projection theorem will turn out to be of extreme importance in carrying out the arguments.

The paper proceeds as follows: The next Section contains definitions and some mathematical preliminaries. The main Bayesian equilibrium existence theorem and its proof are contained in Section 3. Finally, a comparison with the related literature and some remarks are given in Section 4.

2. Mathematical preliminaries

2.1 Notation

2^A denotes the set of all nonempty subsets of the set A.
$\setminus$ denotes the set theoretic subtraction.
$\Re^\ell$ denotes the ℓ-fold Cartesian product of the set of real numbers $\Re$.
$\emptyset$ denotes the empty set.
A^c denotes the complement of the set A.
clA denotes the closure of the set A.

2.2 Definitions

Let X and Y be sets. The graph G_ϕ of a correspondence $\phi : X \to 2^Y$ is the set $G_\phi = \{(x, y) \in X \times Y : y \in \phi(x)\}$. If X and Y are topological spaces, $\phi : X \to 2^Y$ is said to be *lower-semicontinuous* (l.s.c) if the set $\{x \in X : \phi(x) \cap V \neq \emptyset\}$ is open in X for every open subset V of Y; $\phi : X \to 2^Y$ is said to be *upper-semicontinuous* (u.s.c) if the set $\{x \in X : \phi(x) \subset V\}$ is open in X for every open subset V of Y.

If (X, α) and (Y, β) are measurable spaces and $\phi : X \to 2^Y$ is a correspondence, ϕ is said to have a *measurable graph* if G_ϕ belongs to the product σ-algebra $a \otimes \beta$. We are often interested in the situation where (X, α) is a measurable space, Y is a topological space and $\beta = \beta(Y)$ is the Borel σ-algebra of Y. For a correspondence ϕ from a measurable space into a topological space, if we say that ϕ has a measurable graph, it is understood that the topological space is endowed with its Borel σ-algebra (unless specified otherwise). In the same setting as above, i.e., (X, α), a measurable space and Y a topological space, ϕ is said to be *lower measurable* if $\{x \in X : \phi(x) \cap V \neq \emptyset\} \in \alpha$ for every V open in Y.

We now define the notion of a Bochner integrable function. We will follow closely Diestel-Uhl (1977). Let (T, τ, μ) be a finite measure space and X be a Banach space. A function $f : T \to X$ is called *simple* if there exist $x_1, x_2, \ldots, x_n$ in X and $\alpha_1, \alpha_2, \ldots, \alpha_n$ in τ such that $f = \sum_{i=1}^{n} x_i \chi_{\alpha_i}$, where $\chi_{\alpha_i}(t) = 1$ if $t \in \alpha_i$ and $\chi_{\alpha_i}(t) = 0$ if $t \notin \alpha_i$. A function $f : T \to X$ is said to be *μ-measurable* if there exists a sequence of simple functions $f_n : T \to X$ such that $\lim_{n\to\infty} \| f_n(t) - f(t) \| = 0$ for almost all $t \in T$. A μ-measurable function $f : T \to X$ is said to be *Bochner integrable* if there exists a sequence of simple functions $\{f_n : n = 1, 2, \ldots,\}$ such that

$$\lim_{n\to\infty} \int_T \| f_n(t) - f(t) \| \, d\mu(t) = 0.$$

In this case we define for each $E \in \tau$ the *integral* to be $\int_E f(t)\, d\mu(t) = \lim_{n\to\infty} \int_E f_n(t)\, d\mu(t)$. It can be shown [see Diestel-Uhl (1977), Theorem 2, p.45] that, if $\phi : T \to X$ is a μ-measurable function then f is *Bochner integrable* if and only if $\int_T \| f(t) \| \, d\mu(t) < \infty$. It is important to note that the *Dominated Convergence Theorem* holds for Bochner integrable functions, in particular, if $f_n : T \to X \ (n = 1, 2, \ldots)$ is a sequence of Bochner integrable functions such that $\lim_{n\to\infty} f_n(t) = f(t)$ μ-a.e., and $\| f_n(t) \| \leq g(t)$ μ-a.e., where $g \in L_1(\mu, \Re)$, then f is Bochner integrable and $\lim_{n\to\infty} \int_T \| f_n(t) - f(t) \| \, d\mu(t) = 0$.

For $1 \leq p < \infty$, we denote by $L_p(\mu, X)$ the space of equivalence classes of X-valued Bochner integrable functions $x : T \to X$ normed by

$$\| x \|_p = \left(\int_T \| x(t) \|_p \, d\mu(t) \right)^{1/p}.$$

It is a standard result that normed by the functional $\| \cdot \|_p$ above, $L_p(\mu, X)$ becomes a Banach space [see Diestel-Uhl (1977), p.50]. We denote by S_ϕ^p the *set of all selections from* $\phi : T \to 2^X$ that belong to the space $L_p(\mu, X)$, i.e.,

$$S_\phi^p = \{x \in L_p(\mu, X) : x(t) \in \phi(t) \ \ \mu - a.e.\}.$$

We will also consider the set $S^1_\phi = \{x \in L_1(\mu, X) : x(t) \in \phi(t) \;\; \mu - a.e.\}$, i.e., S^1_ϕ is the *set of all Bochner integrable selections from* $\phi(\cdot)$. Recall that the correspondence $\phi : T \to 2^X$ is said to be *integrably bounded* if there exists a map $h \in L_1(\mu, \Re)$ such that $\sup\{\| x \| : x \in \phi(t)\} \leq h(t) \;\; \mu - a.e.$ Moreover, note that if T is a complete measure space, X is a separable Banach space and $\phi : T \to 2^X$ is an integrably bounded, nonempty valued correspondence having a measurable graph, then by virtue of the Aumann measurable selection theorem (see the next Section) we can conclude that S^1_ϕ is nonempty.

2.3 Preliminary theorems

The results below will be useful for the proof of our main theorem. We refer the reader to Yannelis [1991a, 1991b] for more details and further references..

Aumann measurable selection theorem. *Let* (T, τ, μ) *be a complete finite measure space,* Y *be a complete, separable metric space and* $\phi : T \to 2^Y$ *be a nonempty valued correspondence with a measurable graph, i.e.,* $G_\phi \in \tau \otimes \beta(Y)$*. Then there is a measurable function* $f : T \to Y$ *such that* $f(t) \in \phi(t) \;\; \mu - a.e.$

Diestel's theorem. *Let* (T, τ, μ) *be a complete finite measure space,* X *be a separable Banach space and* $\phi : T \to 2^X$ *be an integrally bounded, convex, weakly compact and nonempty valued correspondence. Then* S^1_ϕ *is weakly compact in* $L_1(\mu, X)$.

Projection theorem. *Let* $(\Omega, \mathcal{F}, \mu)$ *be a complete finite measure space and* Y *be a complete, separable metric space. If* H *belongs to* $\mathcal{F} \otimes \beta(Y)$*, its projection* $Proj_\Omega(H)$ *belongs to* $\mathcal{F}$.

It follows from the projection theorem that if a correspondence has a measurable graph then it is also lower measurable. The reverse is also true if the correspondence is closed valued (see Castaing-Valadier (1977, p. 80, Theorem III.30]).

The following Lemma (see also Yannelis [1991b, Theorem 5.4] for a similar argument) will be of fundamental importance for the proof of the main theorem in the next Section.

Lemma 2.1. *Let* Y *be a separable space,* $(\Omega, \mathcal{F}, \mu)$ *be a complete finite measure space and* $X : \Omega \to 2^Y$ *be an integrably bounded, nonempty, convex valued correspondence such that for all* $\omega \in \Omega, X(\omega)$ *is a weakly* *compact, convex subset of Y. Denote by L_X the set $\{x \in L_1(\mu, Y) : x(\omega) \in X(\omega) \ \mu - a.e.\}$. Let $\phi : \Omega \times L_X \to 2^Y$ be a nonempty, closed, convex valued correspondence such that $\phi(\omega, x) \subset X(\omega)$ for all $(\omega, x) \in \Omega \times L_X$. Assume that for each fixed $x \in L_X, \phi(\cdot, x)$ has a measurable graph and that for each fixed $\omega \in \Omega$, $\phi(\omega, \cdot) : L_X \to 2^Y$ is u.s.c. in the sense that the set $\{x \in L_X : \phi(\omega, x) \subset V\}$ is weakly open in L_X for every norm open subset V of Y. Define the correspondence $\Phi : L_X \to 2^{L_X}$ by*

$$\Phi(x) = \{y \in L_X : y(\omega) \in \phi(\omega, x) \ \ \mu - a.e.\}.$$

Then Φ is weakly u.s.c., i.e., the set $\{x \in L_X : \Phi(x) \subset V\}$ is weakly open in L_X for every weakly open subset V of L_X.

Proof. By Diestel's Theorem L_X as well as $\Phi(x)$ for each $x \in L_X$, endowed with the weak topology is compact. Since the weak topology of a weakly compact subset of a separable Banach space is metrizable (Dunford and Schwartz [1958, p.434]), L_X is a compact, metrizable space. Thus, it suffices to show that if x and $\{x_n : n = 1, 2, \dots\}$ belong to $L_X, \{x_n : n = 1, 2, \dots\}$ converges weakly to x, and V is a relatively weakly open subset of L_X containing $\Phi(x)$, then $\Phi(x_n) \subset V$ for all sufficiently large n. For if $U = \{z \in L_X : \Phi(z) \subset V\}$ is not relatively weakly open in L_X, we can pick some $x \in U$ such that every neighborhood of x in the (relative) weak topology of L_X contains an $\tilde{x} \notin U$. Hence, we can construct a sequence $\{x_n\}$ converging weakly to x such that $x_n \in L_X$ and $\Phi(x_n) \not\subset V$.

Let $\mathcal{B}$ and $\tilde{\mathcal{B}}$ denote the closed unit balls in Y and L_X, respectively and let $\varepsilon > 0$. It will suffice to show that for a suitable $n_0, \Phi(x_n) \subset \Phi(x) + \varepsilon\tilde{\mathcal{B}}$ for $n \geq n_0$. This is so, because every weakly open neighborhood of the weakly compact set $\phi(x)$ contains the norm neighborhood $\Phi(x) + \varepsilon\tilde{\mathcal{B}}$ for a suitable $\varepsilon > 0$.

We begin in finding the suitable n_0. Since for each $\omega \in \Omega, \phi(\omega, \cdot)$ is u.s.c, we can find a minimal N_ω such that

$$\phi(\omega, x_n) \subset \Phi(\omega, x) + \frac{\varepsilon}{3\mu(\Omega)\mathcal{B}} \qquad \text{for all } n \geq N_\omega. \tag{3.1}$$

Let $\varepsilon/3\mu(\Omega) = \delta_1$. By assumption, for fixed x and n, the correspondences $\phi(\cdot, x) : \Omega \to 2^Y$ and $\phi(\cdot, x_n) : \Omega \to 2^Y$ have measurable graphs. Thus by the projection theorem the set $\mathcal{Q}$ defined below belongs to $\mathcal{F}$:

$$\mathcal{Q} = proj_\Omega\{(\omega, y) \in \Omega \times Y : (\omega, y) \in G_{\phi(\cdot, x_n)} \cap (G_{\phi(\cdot, x) + \delta_1 \mathcal{B}})^c\} \in \mathcal{F}.$$

Note that

$$\begin{aligned} \mathcal{Q} &= \{\omega \in \Omega : \phi(\omega, x_n) \not\subset \phi(\omega, x) + \delta_1\mathcal{B}\} \\ &= \{\omega \in \Omega : \phi(\omega, x_n) \setminus \phi(\omega, x) + \delta_1\mathcal{B}) \neq \emptyset\}. \end{aligned}$$

This will enable us to conclude that N_ω is a measurable function of ω. This is clearly so, since,

$$\begin{aligned}\{\omega \in \Omega : N_\omega = m\} &= \bigcap_{n \geq m} \{\omega \in \Omega : \phi(\omega, x_n) \subset \phi(\omega, x) + \delta_1 \mathcal{B}\} \\ &\cap \{\omega \in \Omega : \phi(\omega, x_{m-1}) \not\subset \phi(\omega, x) + \delta_1 \mathcal{B}\}.\end{aligned}$$

We are now ready to choose the suitable n_0. Since $X(\cdot)$ is integrably bounded, there exists $g \in L_1(\mu, \Re)$ such that $\sup\{\| x \| : x \in X(\omega)\} \leq g(\omega)$. Pick δ_2 such that if $\mu(A) < \delta_2$, $(A \subset \Omega)$ then $\int_A g(\omega) d\mu(\omega) < \varepsilon/3$. Since N_ω is a measurable function of ω, we can choose n_0 such that $\mu(\{\omega \in \Omega : N_\omega \geq n_0\}) < \delta_2$. This is the desired n_0.

Let $n \geq n_0$ and $y \in \Phi(x_n)$. We must show that $y \in \Phi(x) + \varepsilon \tilde{\mathcal{B}}$.

Since $\phi(\cdot, x)$ has a measurable graph and it is nonempty valued by the Aumann measurable selection theorem, there is a measurable selection $z_1 : \Omega \to Y, z_1(\omega) \in \phi(\omega, x)$ $\mu - a.e.$ in Ω.

The correspondence

$$\psi(\omega) = (\{y(\omega)\} + \delta_1 \mathcal{B}) \cap \phi(\omega, x)$$

has a measurable graph and by (3.1) is nonempty valued for $\omega \in \Omega_0 = \{\omega : N_\omega \leq n_0\}$.

Thus by the Aumann measurable selection theorem, there is a measurable function $z_2 : \Omega \to Y$ such that $z_2(\omega) \in \psi(\omega)$ $\mu - a.e.$ in Ω_0. Let,

$$z(\omega) = z_1(\omega) \quad \text{for } \omega \notin \Omega_0 \qquad \text{and} \qquad z(\omega) = z_2(\omega) \quad \text{for } \omega \in \Omega_0.$$

Then $z(\omega) \in \phi(\omega, x)$ $\mu - a.e.$ and thus $z \in \Phi(x)$. If we show that $\| z - y \| < \varepsilon$, the proof of the Lemma would be complete.

However, this is easy to check. Indeed,

$$\begin{aligned}\| z - y \| &= \int_{\Omega \setminus \Omega_0} \| z_1(\omega) - y(\omega) \| d\mu(\omega) + \int_{\Omega_0} \| z_2(\omega) - y(\omega) \| d\mu(\omega) \\ &< 2 \int_{\Omega \setminus \Omega_0} g(\omega) d\mu(\omega) + \int_{\Omega_0} \delta_1 d\mu(\omega) \\ &< \frac{2}{3}\varepsilon + \delta_1 \mu(\Omega) = \frac{2}{3}\varepsilon + \frac{\varepsilon}{3\mu(\Omega)} \cdot \mu(\Omega) = \varepsilon.\end{aligned}$$

The proof of the lemma is now complete.

3. A Bayesian social equilibrium existence theorem

Let $(\Omega, \mathcal{F}, \mu)$ be a complete, finite, separable measure space, where Ω denotes the set of *states* of nature of the world and the σ-algebra $\mathcal{F}$, denotes the set of *events*. Let Y, denote the *strategy* or *commodity* space, where Y is a separable Banach space.

A *Bayesian abstract economy* (or *social system*) *with differential information* is a set $G = \{(X_t, u_t, \mathcal{F}_t, A_t, q_t) : t \in T\}$, where

1. $X_t : \Omega \to 2^Y$ is the *action* (*strategy*) set-valued function of agent t,
2. for each $\omega \in \Omega, u_t(\omega, \cdot) : \prod_{s\in T} X_s(\omega) \to \Re$ is the *random utility function* of agent t,
3. $\mathcal{F}_t$ is a sub σ-algebra of $\mathcal{F}$ which denotes the *private information* of agent t,
4. for each $\omega \in \Omega, A_t(\omega, \cdot) : \prod_{s\in T} X_s(\omega) \to 2^Y$ is the *random constraint set-valued function* of agent t, where for all $(\omega, x) \in \Omega \times \prod_{t\in T} X_t(\omega)$, $A_t(\omega, x) \subset X_t(\omega)$,
5. $q_t : \Omega \to \Re_{++}$ is the *prior* of agent t, which is a Radon-Nikodym derivative such that $\int_\Omega q_t(\omega) d\mu(\omega) = 1$.

Let $L_{X_t} = \{x \in L_1(\mu, Y) : x_t : \Omega \to Y$ is $\mathcal{F}_t$-measurable and $x_t(\omega) \in X_t(\omega)\ \mu - a.e.\}$. Denote by L_X the set $\prod_{t\in T} L_{X_t}$ and by $L_{X_{-t}}$ the set $\prod_{s\neq t} L_{X_s}$. An element x_t of L_{X_t} is called a *strategy* for player (agent) t. The typical element of L_{X_t} is denoted by $\tilde{x}_t$ and that of $X_t(\omega)$ by $x_t(\omega)$ (or x_t). We assume that for each $t \in T$, there exists a countable partition $\prod_t$ of Ω. Moreover, the σ-algebra $\mathcal{F}_t$ is generated by $\prod_t$. For each $\omega \in \Omega$, let $E_t(\omega) \in \prod_t$ denote the smallest set in $\mathcal{F}_t$ containing ω and assume that for each t,

$$\int_{\omega' \in E_t(\omega)} q_t(\omega') d\mu(\omega') > 0.$$

The *interim expected utility* of agent t, $V_t(\omega, \cdot, \cdot) : L_{X_{-t}} \times X_t(\omega) \to \Re$ is defined as

$$V_t(\omega, \tilde{x}_{-t}, x_t) = \int_{\omega' \in E_t(\omega)} u_t(\omega', \tilde{x}_{-t}(\omega'), x_t) q_t(\omega' | E_t(\omega)) d\mu(\omega')$$

where

$$q_t(\omega' | E_t(\omega)) = \begin{cases} 0 & \text{if } \omega' \notin E_t(\omega) \\ \frac{q_t(\omega')}{\int_{\bar{\omega} \in E_t(\omega)} q_t(\bar{\omega}) d\mu(\bar{\omega})} & \text{if } \omega' \in E_t(\omega). \end{cases}$$

The function $V_t(\omega, \tilde{x}_{-t}, x_t)$ is interpreted as the (interim) *expected utility* of agent t, using his own action (*strategy*) x_t, where the state of nature is ω and the other agents employ the strategy profile $\tilde{x}_{-t}$.

A *Bayesian equilibrium* for G is a strategy profile $\tilde{x}^* \in L_X$ such that for all $t \in T$,

(i) $x_t^*(\omega) \in A_t(\omega, \tilde{x}^*)$ $\mu - a.e.$

(ii) $V_t(\omega, \tilde{x}^*_{-t}, \tilde{x}_t^*(\omega)) = \max_{y_t \in A_t(\omega, \tilde{x}^*)} V_t(\omega, \tilde{x}^*_{-t}, y_t)$ $\mu - a.e.$

Assumptions

(A.1)

(a) $X_t : \Omega \to 2^Y$ is a nonempty, convex, weakly compact-valued and integrably bounded correspondence.

(b) $X_t : \Omega \to 2^Y$ is $\mathcal{F}_t$-lower measurable, i.e., for every open subset V of Y, the set, $\{\omega \in \Omega : X_t(\omega) \cap V \neq \phi\}$ belongs to $\mathcal{F}_t$.[1]

(A.2)

(a) For each fixed $x, u_t(\cdot, x) : \Omega \to \Re$ is $\mathcal{F}$-measurable.

(b) For each $\omega \in \Omega$ and $x_{-t} \in \prod_{s \neq t} X_s(\omega), u_t(\omega, x_{-t}, \cdot) : X_t(\omega) \to \Re$ is concave.

(c) u_t is integrably bounded.

(d) For each $\omega \in \Omega, u_t(\omega, \cdot, \cdot) : \prod_{s \neq t} X_s(\omega) \times X_t(\omega) \to \Re$ is continuous where $X_s(\omega), s \neq t$, is endowed with the weak topology and $X_t(\omega)$ with the norm topology.

(A.3)

(a) For each $(\omega, x) \in \Omega \times \prod_{t \in T} X_t(\omega), A_t(\omega, x)$ is lower measurable, i.e., for every open subset V of Y, the set $\{(\omega, x) \in \Omega \times \prod_{t \in T} X_t(\omega) : A_t(\omega, x) \cap V \neq \phi\}$ belongs to $\mathcal{F} \otimes \beta(\prod_{t \in T} X_t(\omega))$.

(b) For each $\omega \in \Omega, A_t(\omega, \cdot) : \prod_{t \in T} X_t(\omega) \to 2^Y$ is a weakly continuous correspondence with closed, convex, nonempty values.

Theorem 3.1. *Let T be a countable set. Let $G = \{(X_t, u_t, A_t, \mathcal{F}_t, q_t) : t \in T\}$ be a Bayesian abstract econom satisfying (A.1) - (A.3). Then there exists a Bayesian equilibrium for G.*

Proof. It follows from the Lebesgue dominated convergence theorem that $V_t(\omega, \cdot, \cdot) : L_{X_{-t}} \times X_t(\omega) \to \Re$ is continuous where $L_{X_s}, (s \neq t)$ is endowed with the weak topology and $X_t(\omega)$ with the norm topology, (see for example Lemma A.1, p.345 in Kim-Yannelis [1997]). Moreover, for each $(\tilde{x}_{-t}, x_t) \in L_{X_{-t}} \times X_t(\omega)$, $V_t(\cdot, \tilde{x}_{-t}, x_t) : \Omega \to \Re$ is $\mathcal{F}_t$-measurable and from the concavity of u_t in x_t it follows that V_t is also concave in x_t.

For each $t \in T$, define $\bar{A}_t : \Omega \times L_X \to 2^Y$ by $\bar{A}_t(\omega, \tilde{x}) = A_t(\omega, x)$. For each $t \in T$, define the correspondence

[1] Notice that in the presence of assumption A.1(a) the $\mathcal{F}_t$-lower measurability of $X_t(\cdot)$ is equivalent to the $\mathcal{F}_t$-measurable graph of $X_t(\cdot)$. This is also the case for the lower measurability assumption of the random constraint correspondence (assumption, A.3(a), below).

$\phi_t : \Omega \times L_{X_{-t}} \to 2^Y$ by

$$\phi_t(\omega, \tilde{x}_{-t}) = \{x_t \in \bar{A}_t(\omega, \tilde{x}) : V_t(\omega, \tilde{x}_{-t}, x_t) = \max_{y_t \in \bar{A}_t(\omega,\tilde{x})} V_t(\omega, \tilde{x}_{-t}, y_t)\}$$

It is easy to see that ϕ_t is nonempty valued. Indeed, since all the values of A_t are contained in the weakly compact valued correspondence X_t and A_t is (norm) closed and convex, so weakly closed, we can conclude that $\bar{A}_t$ is weakly compact valued. Since V_t is norm u.s.c. in $X_t(\omega)$ and also concave, it is weakly u.s.c. (Balder-Yannelis [1993, p. 629]) and therefore we can conclude that ϕ_t is nonempty valued. Moreover, it follows from Berge's maximum theorem that for each fixed $\omega \in \Omega$, $\phi_t(\omega, \cdot)$ is u.s.c. in the sense that the set $\{x_{-t} \in L_{X_{-t}} : \phi_t(\omega, x_{-t}) \subset V\}$ is weakly open in $L_{X_{-t}}$ for every norm open subset V of Y.

For each $t \in T$, define $\Phi_t : L_{X_{-t}} \to 2^{L_{X_t}}$ by

$$\Phi_t(\tilde{x}_{-t}) = \{\tilde{x}_t \in L_{X_t} : \tilde{x}_t(\omega) \in \phi_t(\omega, \tilde{x}_{-t}) \ \ \mu - a.e.\}$$

We must first show that Φ_t is nonempty valued. It follows from Lemma III.39 in Castaing-Valadier [1977] that for each fixed $\tilde{x}_{-t}$, $\phi_t(\cdot, \tilde{x}_{-t})$ has a measurable graph. Moreover, ϕ_t is nonempty valued. Hence, by the Aumann measurable selection theorem for each fixed $\tilde{x}_{-t} \in L_{X_{-t}}$ there exists an $\mathcal{F}_t$-measurable function $f_t : \Omega \to Y$ such that $f_t(\omega) \in \phi_t(\omega, \tilde{x}_{-t})$ $\mu - a.e.$. Since for each $(\omega, \tilde{x}_{-t}) \in \Omega \times L_{X_{-t}}$, $\phi_t(\omega, \tilde{x}_{-t})$ is contained in the integrably bounded correspondence $X_t(\cdot)$, then $f_t \in L_{X_t}$ and we conclude that $f_t \in \Phi_t(\tilde{x}_{-t})$ for every $\tilde{x}_{-t} \in L_{X_{-t}}$. Thus, Φ_t is nonempty valued and it is clearly convex valued since ϕ_t is so. A similar argument which the above can be followed to show that each L_{X_t} is nonempty and therefore so is L_X. By Diestel's Theorem L_{X_t} is a weakly compact subset of $L_1(\mu, Y)$. By virtue of Theorem 3 in Dunford-Schwartz [1958, p. 434], we conclude that L_{X_t} is metrizable. Since T is a countable set, the set $L_{X_{-t}}$ is also metrizable. Also, $L_{X_{-t}}$ is weakly compact. By Lemma 2.1, Φ_t is weakly u.s.c. Define $\Phi : L_X \to 2^{L_X}$ by

$$\Phi(\tilde{x}) = \prod_{t \in T} \Phi_t(\tilde{x}_{-t}).$$

One can easily check that Φ is also a weakly u.s.c. correspondence with convex, closed, nonempty valued. Since the set L_X is weakly compact, convex and nonempty, by the Fan-Glicksberg fixed point theorem there exists an $\tilde{x}^* \in L_X$ such that $\tilde{x}^* \in \Phi(\tilde{x}^*)$. One can easily verify that $\tilde{x}^*$ is a Bayesian equilibrium for G.

4. Related literature and concluding remarks

4.1 Related literature

First it should be noted that if for each $\omega \in \Omega, E_i(\omega) = \{\omega\}$ then $V_i(\omega, x) = u_i(\omega, x)$ and our theorem reduces to a random equilibrium. Moverover, if for each $\omega \in \Omega$, we set $X_t(\omega) = X_t$, where X_t is a compact, convex, nonempty subset of Y and also set for each $(\omega, x) \in \Omega \times \prod_{t\in T} X_t, A_t(\omega, x) = A_t(x)$ and $u_t(\omega, x) = u_t(x)$ we can deduce from our main result, a generalization of the deterministic existence result of Debreu (1952) for an abstract economy with an infinite dimensional strategy space and with a countable number of players.

It should be also noted that by setting for all $(\omega, x) \in \Omega \times \prod_{t\in T} X_t$, $A_t(\omega, x) = X_t(\omega)$ we obtain as a corollary, Theorem 4.1 of Kim-Yannelis (1997), as well as Theorem 4.1 in Yannelis (1998). Finally, our theorem generalizes Theorem 4.1 in Aliprantis-Tourky-Yannelis (2001) from a finite dimensional strategy space to an infinite dimensional one. Moreover, the infinite dimensional equilibrium result in Aliprantis-Tourky-Yannelis (2001) has a slightly stronger assumption on the random constraint correspondences which imply ours. In particular, in Aliprantis-Tourky-Yannelis it is assumed that:

1. $A_t : \Omega \times \prod_{s\in T} X_s \to 2^Y$ is u.s.c.
2. There exists a correspondence $\tilde{A}_t : \Omega \times \prod_{s\in T} X_s \to 2^Y$ such that:
   (a) $cl\tilde{A}_t(\omega, x) = A_t(\omega, x)$ for all $(\omega, x) \in \Omega \times \prod_{t\in T} X_t$
   (b) for each $\omega \in \Omega, \tilde{A}_t(\omega, \cdot)$ has open lower sections and it is convex, nonempty valued.

In view of Proposition 3.1 in Yannelis (1987) we can easily see that assumptions (1) and (2) above imply A.3(b), but the reverse is not true.

Finally, it is important to point out that the method of proof in Aliprantis-Tourky-Yannelis (ATY) is different that the one in this paper. In particular, ATY make use of Caratheodory-type selections. To be more precise, the present proof follows the footsteps of the one of Nash (1950)[2] and Debreu (1952) which relies on the fixed point of the best reply correspondence. However, the ATY follows the Yannelis-Prabhakar (1983) approach which relys on a continuous selection theorem. Of course, the mathematical argument needed to provide Bayesian generalizations following either approach are far from trivial.

[2] Nash (1950) attributes the method of proof (fixed point of the best reply mapping) to David Gale. The best reply mapping proof simplifies the original one of Nash (1951) which makes use of an ingenious mapping on which the Brouwer fixed point theorem is applied.

4.2 Concluding remarks

1. Notice that in the definition of a Bayesian equilibrium for the Bayesian abstract economy G, the interim expected utility was used. One may want to consider the *ex ante utility function* of agent t, defined as,

$$V_t(\tilde{x}_{-t}, x_t) = \int_{\Omega} u_t(\omega, \tilde{x}_{-t}(\omega), x_t(\omega)) d\mu(\omega).$$

   A Bayesian equilibrium can be defined in terms of the ex ante expected utility and the proof of our theorem covers this case as well.
2. The social equilibrium existence theorem of Debreu, was the main mathematical tool to prove the existence of a Walrasian equilibrium as shown by Arrow-Debreu (1954). In a similar fashion, one can show that the abstract Bayesian equilibrium theorems can be used to prove the existence of an expectations (Radner) equilibrium for an economy with differential information (see for example Radner (1968) or Einy-Moreno-Shitovitz (2001), among others). We hope to take there details up in a subsequent paper.
3. By imposing a measurability assumption for a mapping from the set of agent to the strategy set (e.g., Kim-Yannelis [1997, B.4, p. 342]) we can employ a similar argument with the one in this paper to extend our existence result to a measure space of agents (see also Balder (2001) for such results).
4. In Section 3 the random constraint set valued function was defined as a mapping from $\Omega \times \prod_{t \in T} X_t(\omega)$ to 2^Y. The Theorem 3.1 and its proof remain unchanged if the random constraint correspondence is defined as, $A_t : \Omega \times L_X \to 2^Y$.

References

[1] Aliprantis, C.D., Tourky, R., Yannelis, N.C.: A Bayesian generalization of Debreu's social equilibrium existence theorem. (manuscript 2001)
[2] Arrow, J., Debreu, G.: Existence of an equilibrium for a competitive economy. Econometrica **22**, 265-290 (1954)
[3] Balder, E.J.: A unifying pair of Cournot-Nash equilibrium existence results. (manuscript 2001)
[4] Balder, E.J., Yannelis, N.C.: On the continuity of expected utility. Economic Theory **3**, 625-643 (1993)
[5] Castaing, C., Valadier. M.: Convex Analysis and Measurable Multifunctions. Lecture Notes in Mathematics, Vol. 580, Springer-Verlag, New York 1997

[6] Debreu, G.: A social equilibrium existence theorem. Proc. Natl. Acad. Sci. USA **38**, 886-803 (1952)
[7] Diestel, J., Uhl, J.: Vector Measures, Mathematical Surveys, Vol.15. Amer. Math. Soc., Providence 1977
[8] Dunford, N., Schwartz, J.T.: Linear Operators, Vol. I. Interscience, New York 1958
[9] Einy, E., Moreno, D.,Shitovitz, B.: Competitive and core allocations in large economies with differential information. Economic Theory (2001)
[10] Kim, T., Yannelis, N.C.: Existence of equilibrium in Bayesian games with infinitely many players. Journal of Economic Theory **77**, 330-353 (1977)
[11] Nash, J.F.: Noncooperative games. Ann. Math. **54**, 286-295 (1951)
[12] Nash, J.F.: Equilibrium points in n-person games. Proc.Natl.Acad.Sci.USA **36**, 48-49 (1950)
[13] Radner, R.: Competitive equilibrium under uncertainty. Econometrica **36**, 31-58 (1968)
[14] Yannelis, N.C.: Equilibrium in noncooperative models of competition. Journal of Economic Theory **41**, 96-111 (1987)
[15] Yannelis, N.C., Prabhakar, N.D.: Existence of maximal elements and equilibria in linear topological spaces. J. Math. Econ. **12**, 233-245 (1983)
[16] Yannelis, N.C.: Set-valued function of two variables in economic theory. In: Khan, M.A., Yannelis, N.C. eds., Equilibrium Theory in Infinite Dimensional Spaces (Studies in Economic Theory). Springer-Verlag 1991a
[17] Yannelis, N.C.: Integration of Banach-valued correspondences. In: Khan, M.A., Yannelis, N.C. eds., Equilibrium Theory in Infinite Dimensional Spaces (Studies in Economic Theory). Springer-Verlag 1991b
[18] Yannelis, N.C.: On the Existence of Bayesian Nash Equilibrium. In: Agramovich, Y., Avgerinos, E., Yannelis, N.C. eds., Functional Analysis and Economic Theory, Springer-Verlag 1998

Subject Index

Instructions for Authors

A. General

1. Papers submitted for publication will be considered only if they have not been and will not be published elsewhere without permission from the publisher and the Research Center of Mathematical Economics.

2. Every submitted paper will be subject to review. The names of reviewers will not be disclosed to the authors or to anybody not involved in the editorial process.

3. The authors are asked to transfer the copyright to their articles to Springer-Verlag if and when these are accepted for publication.

The copyright covers the exclusive and unlimited rights to reproduce and distribute the article in any form of reproduction. It also covers translation rights for all languages and countries.

4. Manuscript must be written in English. One original and 3 sets of photocopies should be submitted to the following address.

Professor Toru Maruyama
Department of Economics
Keio University
2-15-45 Mita, Minato-ku, Tokyo
108-8345 Japan

5. **Offprints**: Twenty-five (25) **offprints** of each paper will be supplied free of charge. Additional offprints can be ordered at cost price.

B. Preparation of Manuscript

1. Manuscripts must be double-spaced (not 1.5), with wide margins (at least 25 mm), and large type (at least 12 point) on one side of A4 paper. Any manuscript that does not meet these requirements will be returned to the author immediately for retyping.

2. All manuscripts would finally be composed using our Springer LaTeX macro package. The Springer macros use standard LaTeX packages - if you do not have these packages refer to www.dante.de/cgi-bin/ctan-index. If authors make manuscripts by word-processing software other than TeX, please follow our styles as possible. For authors who prepare their manuscripts by TeX, we strongly recommend to visit our homepage, http://www.springer.de/economics/authors/interinf.html, where you can download all the necessary macro packages with instructions for LaTeX2e, LaTeX 2.09 and plain tex. For support, please contact `texhelp@springer.de`.

3. **The title page** must include: title; short (running) title of up to 60/85 characters; first names and surnames of all coauthors with superscript numerals indicating their affiliations; full street addresses of all affiliations; address to which proofs should be sent; fax number; and any footnotes referring to the title (indicated by asterisks*).

4. **Summary/abstract**: Each paper must be preceded by a summary/an abstract, which should not exceed 100 words.

5. **The Journal of Economic Literature index number (JEL classification)** should be indicated and the statement of the **2000 Mathematics Subject Classification (MSC) numbers** is desirable. You can check JEL classification with Internet at `http://ideas.uqam.ca/ideas/data/JEL` as well as 2000 MSC numbers at `http://e-math.ams.org/msc`.

6. **Main text**: All tables and figures must be cited in the text and numbered consecutively with Arabic numerals according to the sequence in which they are cited. Please mark the desired position of tables and figures in the left margin of the manuscript.

Do not italicize, underscore, or use boldface for headings and subheadings. Words that are to be italicized should be underscored in pencil.

Abbreviations must be spelled out at first mention in summary/abstract and main text. Abbreviations should also be spelled out at first mention in figure legends and table footnotes.

Short equations can be run in with the text. Equations that are displayed on a separate line should be numbered.

7. **References**: The list of references should be in alphabetical order and include the names and initials of all authors (see examples below). Whenever possible, please update all references to papers accepted for publication, preprints or technical reports, giving the exact name of the journal, as well as the volume, first and last page numbers and year, if the article has already been published or accepted for publication.

When styling the references, the following examples should be observed:

Journal article:
1. or [F-M] Freed, D.S., Melrose, R.B.: A mod *k* index theorem. Invent. Math. **107**, 283-299 (1992)

Complete book:
2. or [C-S] Conway, J.H., Sloane, N.J.: Sphere packings, lattices, and groups (Grundlehren Math. Wiss. Bd. 290) Berlin Heidelberg New York: Springer 1988

Single contribution in a book:
3. or [B] Border, K.C.: Functional analytic tools for expected utility theory. In: Aliprantis, C.D. et al. (eds.): Positive operators, Riesz spaces and economics. Berlin Heidelberg New York: Springer 1991, pp. 69-88

8. **Citations in the text** should be either by numbers in square brackets, e.g. [1], referring to an alphabetically ordered and numbered list, or by the author's initials in square brackets, e.g. [F-M], or by author and year in parentheses, e.g. Freed and Melrose (1992). Any of these styles is acceptable if used consistently throughout the paper. In the third system, if a work with more than two authors is cited, only the first author's name plus "et al." need be given and if there is more than one reference by the same author or team of authors in the same year, then a, b, c should be added after the year both in the text and in the list of references.

9. **Tables** are to be numbered separately from the illustrations. Each table should have a brief and informative title. All abbreviations used in a table must be defined in a table footnote on first use, even if already defined in the text. In subsequent tables abbreviations need not be redefined. Individual footnotes should be indicated by superscript lowercase a, b, c, etc. Permission forms must be provided for any tables from previously published sources (same procedure as with figures; see below).

10. **Figures**: If you have access to suitable software, you can design your figures directly on a computer, but creation by other means is of course also possible. In any case the originals should be pasted into the camera- ready copy at the appropriate places and with the correct orientation. If necessary, the figures may be drawn overscale, but in this case suitably reduced copies should be pasted into the script.

If a figure has been published previously, acknowledge its source and submit written permission signed by author and publisher. The source must be included in the reference list. If a permission fee is required, it must be paid by the author. Responsibility for meeting this requirement lies entirely with the author.

MIX
Papier aus verantwortungsvollen Quellen
Paper from responsible sources
FSC® C105338

If you have any concerns about our products,
you can contact us on
ProductSafety@springernature.com

In case Publisher is established outside the EU,
the EU authorized representative is:
Springer Nature Customer Service Center GmbH
Europaplatz 3, 69115 Heidelberg, Germany

Printed by Libri Plureos GmbH
in Hamburg, Germany